FUZZY CLUSTER ANALYSIS

FUZZY CLUSTER ANALYSIS

METHODS FOR CLASSIFICATION, DATA ANALYSIS AND IMAGE RECOGNITION

Frank Höppner
German Aerospace Center, Braunschweig, Germany
Frank Klawonn
University of Ostfriesland, Emden, Germany
Rudolf Kruse
University of Magdeburg, Germany
Thomas Runkler
Siemens AG, Munich, Germany

JOHN WILEY & SONS, LTD
Chichester • New York • Weinheim • Brisbane • Singapore • Toronto

Originally published in the German language by Friedr. Vieweg & Sohn Verlagsgesellschaft mbH, D-65189 Wiesbaden, Germany, under the title "Frank Höppner/Frank Klawonn/Rudolf Kruse: Fuzzy-Clusteranalysen. 1. Auflage (1st Edition)". Copyright 1997 by Friedr. Vieweg & Sohn Verlagsgesellschaft mbH, Braunschweig/Wiesbaden.

Reprinted January 2000

Copyright © 1999 John Wiley & Sons Ltd
                  Baffins Lane, Chichester,
                  West Sussex, PO19 1UD, England

                  National   01243 779777
                  International (+44) 1243 779777

e-mail (for orders and customer service enquiries): cs-books@wiley.co.uk

Visit our Home Page on http://www.wiley.co.uk or http://www.wiley.com

Other Wiley Editorial Offices

John Wiley & Sons, Inc., 605 Third Avenue,
New York, NY 10158-0012, USA

Weinheim • Brisbane • Singapore • Toronto

Library of Congress Cataloging-in-Publication Data

Fuzzy-Clusteranalysen. English
   Fuzzy cluster analysis : methods for classification, data
analysis, and image recognition / Frank Höppner ... [et al.].
     p.   cm.
   Includes bibliographical references and index.
   ISBN 0-471-98864-2 (cloth : alk. paper)
   1. Cluster analysis. 2. Fuzzy sets. I. Höppner, Frank.
  II. Title.
  QA278.F8913   1999   99-25473
  519.5'3—dc21   CIP

British Library Cataloguing in Publication Data

A catalogue record for this book is available from the British Library

ISBN 0 471 98864 2

Produced from camera-ready copy supplied by the authors
Printed and bound by Antony Rowe Ltd, Eastbourne
This book is printed on acid-free paper responsibly manufactured from sustainable forestry
in which at least two trees are planted for each one used for paper production.

Contents

Preface

When Lotfi Zadeh introduced the notion of a "fuzzy set" in 1965, his primary objective was to set up a formal framework for the representation and management of vague and uncertain knowledge. More than 20 years passed until fuzzy systems became established in industrial applications to a larger extent. Today, they are routinely applied especially in the field of control engineering. As a result of their success to translate knowledge-based approaches into a formal model that is also easy to implement, a great variety of methods for the usage of fuzzy techniques has been developed during the last years in the area of data analysis. Besides the possibility to take into account uncertainties within data, fuzzy data analysis allows us to learn a transparent and knowledge-based representation of the information inherent in the data. Areas of application for fuzzy cluster analysis include exploratory data analysis for pre-structuring data, classification and approximation problems, and the recognition of geometrical shapes in image processing.

When writing this book, our intention was to give a self-contained and methodical introduction to fuzzy cluster analysis with its areas of application and to provide a systematic description of different fuzzy clustering techniques, from which the user can choose the methods appropriate for his problem. The book applies to computer scientists, engineers and mathematicians in industry, research and teaching, who are occupied with data analysis, pattern recognition or image processing, or who take into consideration the application of fuzzy clustering methods in their area of work. Some basic knowledge in linear algebra is presupposed for the comprehension of the techniques and especially their derivation. Familiarity with fuzzy systems is not a requirement, because only in the chapter on rule generation with fuzzy clustering, more than the notion of a "fuzzy set" is necessary for understanding, and in addition, the basics of fuzzy systems are provided in that chapter.

Although this title is presented as a text book we have not included exercises for students, since it would not make sense to carry out the algorithms by hand. We think that applying the algorithms to example data sets is the appropriate way to get a better understanding of the techniques. A software tool implementing most of the algorithms presented in chapters 1–5 and 7 together with the many example data sets discussed in this book are available as public domain software via the Internet at http://fuzzy.cs.uni-magdeburg.de/clusterbook/.

The book is an extension of a translation of our German book on fuzzy cluster analysis published by Vieweg Verlag in 1997. Most parts of the translation were carried out by Mark-Andre Krogel. The book would probably have appeared years later without his valuable support. The material of the book is partly based on lectures on fuzzy systems, fuzzy data analysis and fuzzy control that we gave at the Technical University of Braunschweig, at the University "Otto von Guericke" Magdeburg, at the University "Johannes Kepler" Linz, and at Ostfriesland University of Applied Sciences in Emden. The book is also based on a project in the framework of a research contract with Fraunhofer-Gesellschaft, on results from several industrial projects at Siemens Corporate Technology (Munich), and on joint work with Jim Bezdek at the University of West Florida. We thank Wilfried Euing and Hartmut Wolff for their advisory support during this project.

We would also like to express our thanks for the great support to Juliet Booker, Rhoswen Cowell, Peter Mitchell from Wiley and Reinald Klockenbusch from our German publisher Vieweg Verlag.

Frank Höppner
Frank Klawonn
Rudolf Kruse
Thomas Runkler

Introduction

For a fraction of a second, the receptors are fed with half a million items of data. Without any measurable time delay, those data items are evaluated and analysed, and their essential contents are recognized.

A glance at an image from TV or a newspaper, human beings are capable of this technically complex performance, which has not yet been achieved by any computer with comparable results. The bottleneck is no longer the optical sensors or data transmission, but the analysis and extraction of essential information. A single glance is sufficient for humans to identify circles and straight lines in accumulations of points and to produce an assignment between objects and points in the picture. Those points cannot always be assigned unambiguously to picture objects, although that hardly impairs human recognition performance. However, it is a big problem to model this decision with the help of an algorithm. The demand for an automatic analysis is high, though. Be it for the development of an autopilot for vehicle control, for visual quality control or for comparisons of large amounts of image data. The problem with the development of such a procedure is that humans cannot verbally reproduce their own procedures for image recognition, because it happens unconsciously. Conversely, humans have considerable difficulties recognizing relations in multi-dimensional data records that cannot be graphically represented. Here, they are dependent on computer supported techniques for data analysis, for which it is irrelevant whether the data consists of two- or twelve-dimensional vectors.

The introduction of fuzzy sets by L.A. Zadeh [104] in 1965 defined an object that allows the mathematical modelling of imprecise propositions. Since then this method has been employed in many areas to simulate how inferences are made by humans, or to manage uncertain information. This method can also be applied to data and image analysis.

Cluster analysis deals with the discovery of structures or groupings within data. Since hardly ever any disturbance or noise can be completely eliminated, some inherent data uncertainty cannot be avoided. That is

1

why fuzzy cluster analysis dispenses with unambiguous mapping of the data to classes and clusters, and instead computes degrees of membership that specify to what extend data belong to clusters.

The introductory chapter 1 relates fuzzy cluster analysis to the more general areas of cluster and data analysis, and provides the basic terminology. Here we focus on objective function models whose aim is to assign the data to clusters so that a given objective function is optimized. The objective function assigns a quality or error to each cluster arrangement, based on the distance between the data and the typical representatives of the clusters. We show how the objective function models can be optimized using an alternating optimization algorithm.

Chapter 2 is dedicated to fuzzy cluster analysis algorithms for the recognition of point-like clusters of different size and shape, which play a central role in data analysis.

The linear clustering techniques described in chapter 3 are suitable for the detection of clusters formed like straight lines, planes or hyperplanes, because of the suitable modification of the distance function that occurs in the objective functions. These techniques are appropriate for image processing, as well as for the construction of locally linear models of data with underlying functional interrelations.

Chapter 4 introduces shell clustering techniques, that aim to recognize geometrical contours such as borders of circles and ellipses by further modifications of the distance function. An extension of these techniques to non-smooth structures such as rectangles or other polygons is given in chapter 5.

The cluster estimation models described in chapter 6 abandon the objective function model. This allows handling of complex or not explicitly accessible systems, and leads to a generalized model with user-defined membership functions and prototypes.

Besides the assignment of data to classes, the determination of the number of clusters is a central problem in data analysis, which is also related to the more general problem of cluster validity. The aim of cluster validity is to evaluate whether clusters determined in an analysis are relevant or meaningful, or whether there might be no structure in the data that is covered by the clustering model. Chapter 7 provides an overview on cluster validity, and concentrates mainly on methods to determine the number of clusters, which are tailored to the different clustering algorithms.

Clusters can be interpreted as if-then rules. The structure information discovered by fuzzy clustering can therefore be translated to human readable fuzzy rule bases. The necessary techniques for this rule extraction are presented in chapter 8.

Readers who are interested in watching the algorithms at work can download free software via the Internet from http://fuzzy.cs.uni-magdeburg.de/clusterbook/.

Chapter 1

Basic Concepts

In everyday life, we often find statements like this:

> After a detailed analysis of the data available, we developed the opinion that the sales figures of our product could be increased by including the attribute *fuzzy* in the product's title.

Data analysis is obviously a notion which is readily used in everyday language. Everybody can understand it – however, there are different interpretations depending on the context. This is why these intuitive concepts like data, data analysis, cluster and partition have to be defined first.

1.1 Analysis of data

The concept of a datum is difficult to formalize. It originates from Latin and means "to be given". A datum is arbitrary information that makes an assertion about the state of a system, such as measurements, balances, degrees of popularity or On/Off states. We summarize the totality of all possible states, in which a system can be, under the concept *state* or *data space*. Any element of a data space describes a particular state of a system.

The data that has to be analysed may come from the area of medical diagnosis in the form of a database about patients, they may describe states of an industrial production plant, they may be available as time series which specify the progression of share prices, they may be obtained from statistical investigations, they may reflect opinions of experts, or they may be available as images.

5

Data analysis is always conducted to answer a particular question. That question implicitly determines the form of the answer: although it is dependent on the respective state of the system, it will always be of a particular *type*. Similarly, we want to summarize the possible answers to a question in a set that we call *result space*. In order to really gain information from the analysis, we require the result space to allow at least two different results. Otherwise, the answer would already unambiguously be given without any analysis.

In [5], data analysis is divided into four levels of increasing complexity. The first level consists of a simple frequency analysis, a reliability or credibility evaluation after which data identified as outliers are marked or eliminated, if necessary. On the second level, pattern recognition takes place, by which the data is grouped, and the groups are further structured, etc. These two levels are assigned to the area of exploratory data analysis, which deals with the investigation of data without assuming a mathematical model chosen beforehand that would have to explain the occurrence of the data and their structures. Figure 1.1 shows a set of data where an exploratory data analysis should recognize the two groups or clusters and assign the data to the respective groups.

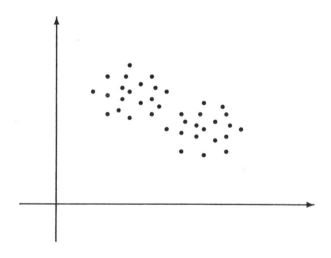

Figure 1.1: Recognition of two clusters by exploratory data analysis

On the third level of data analysis, the data are examined with respect to one or more mathematical models – for example in figure 1.1, whether the assumption is reasonable that the data are realizations of two two-dimensional normally distributed random variables, and if so, what are the underlying parameters of the normal distributions. On the third level, a quantitative data analysis is usually performed, that means (functional) relations between the data should be recognized and specified if necessary, for instance by an approximation of the data using regression. In contrast, a purely qualitative investigation takes place on the second level, with the aim to group the data on the basis of a similarity concept.

Drawing conclusions and evaluating them is carried out on the fourth level. Here, conclusions can be predictions of future or missing data or an assignment to certain structures, for example, which pixels belong to the legs of a chair. An evaluation of the conclusions contains a judgement about how reliably the assignments can be made, whether modelling assumptions are realistic at all, etc. If necessary, a model that was constructed on the third level has to be revised.

The methods of fuzzy cluster analysis introduced in chapter 2 can essentially be categorized in the second level of data analysis, while the generation of fuzzy rules in chapter 8 belongs to the third level, because the rules serve as a description of functional relations. Higher order clustering techniques can also be assigned to the third level. Shell clustering, for example, not only aims at mapping of the data to geometrical contours such as circles, but is also used for a determination of parameters of geometrical contours, such as the circle's centre and radius.

Fuzzy clustering is a part of fuzzy data analysis that comprises two very different areas: the analysis of fuzzy data and the analysis of usual (crisp) data with the help of fuzzy techniques. We restrict ourselves mainly to the analysis of crisp data in the form of real-valued vectors with the help of fuzzy clustering methods. The advantages offered by a fuzzy assignment of data to groups in comparison to a crisp one will be clarified later on.

Even though measurements are usually afected by uncertainty, in most cases they provide concrete values so that fuzzy data are rarely obtained directly. An exception are public opinion polls that permit evaluations such as "very good" or "fairly bad" or, for instance, statements about time aspects such as "for quite a long time" or "for a rather short period of time". Statements like these correspond more to fuzzy sets than crisp values or intervals and should therefore be modelled with fuzzy sets. Methods to analyse fuzzy data like these are described in [6, 69, 73], among others. Another area, where fuzzy data are produced, is image processing. Grey values in grey scale pictures can be interpreted as degrees of membership to the colour black so that a grey scale picture represents a fuzzy set over the pixels. Even though we apply the fuzzy clustering techniques that are introduced in this book for image processing of black-and-white pictures only, these techniques can be extended to grey scale pictures by assigning each pixel its grey value (transformed into the unit interval) as a weight. In this sense, fuzzy clustering techniques especially for image processing can be considered as methods to analyse fuzzy data.

1.2   Cluster analysis

Since the focus lies on fuzzy cluster analysis methods in this book, we can give only a short survey on general issues of cluster analysis. A more thorough treatment of this topic can be found in monographs such as [3, 16, 96].

The aim of a cluster analysis is to partition a given set of data or objects into clusters (subsets, groups, classes). This partition should have the following properties:

- Homogeneity within the clusters, i.e. data that belong to the same cluster should be as similar as possible.

- Heterogeneity between clusters, i.e. data that belong to different clusters should be as different as possible.

The concept of "similarity" has to be specified according to the data. Since the data are in most cases real-valued vectors, the Euclidean distance between data can be used as a measure of the dissimilarity. One should consider that the individual variables (components of the vector) can be of different relevance. In particular, the range of values should be suitably scaled in order to obtain reasonable distance values. Figures 1.2 and 1.3 illustrate this issue with a very simple example. Figure 1.2 shows four data points that can obviously be divided into the two clusters $\{x_1, x_2\}$ and $\{x_3, x_4\}$. In figure 1.3, the same data points are presented using a different scale where the units on the x-axis are closer together while they are more distant on the y-axis. The effect would be even stronger if one would take kilo-units for the x-axis and milli-units for the y-axis. Two clusters can be recognized in figure 1.3, too. However, they combine the data point x_1 with x_4 and x_2 with x_3, respectively.

Further difficulties arise when not only real-valued variables occur but also integer-valued ones or even abstract classes (e.g. types of cars: convertible, sedan, truck etc.). Of course, the Euclidean distance can be computed for integer values. However, the integer values in a variable can produce a cluster partition where a cluster is simply assigned to each occurring integer number. That can be meaningful or completely undesirable dependent on the data and the question to be investigated. Numbers can be assigned to abstract classes, and thus the Euclidean distance can be applied again.

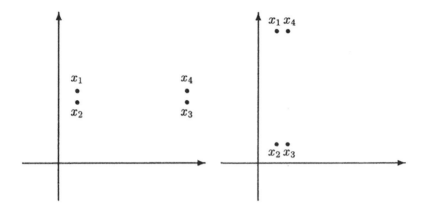

Figure 1.2: Four data points   Figure 1.3: Change of scales

However in this way additional assumptions are used, for example, that the abstract class, which is assigned the number one, is more similar to the second class than to the third.

It would exceed the scope of this book to introduce the numerous methods of classical clustering in detail. Therefore, we present only the main (non-disjoint) families of conventional clustering techniques.

- *incomplete* or *heuristic cluster analysis techniques:* These are geometrical methods, representation or projection techniques. Multi-dimensional data are analysed by dimension reduction such as a principal component analysis (PCA), in order to obtain a graphical representation in two or three dimensions. Clusters are determined subsequently, e.g. by heuristic methods based on the visualization of the data.

- *deterministic crisp cluster analysis techniques:* With these techniques, each datum will be assigned to exactly one cluster so that the cluster partition defines an ordinary partition of the data set.

- *overlapping crisp cluster analysis techniques:* Here, each datum will be assigned to at least one cluster, or it may be simultaneously assigned to several clusters.

- *probabilistic cluster analysis techniques:* For each datum, a probability distribution over the clusters is determined that specifies the probability with which a datum is assigned to a cluster. These techniques are also called fuzzy clustering algorithms if the probabilities are interpreted as degrees of membership.

- *possibilistic cluster analysis techniques:* These techniques are pure fuzzy clustering algorithms. Degrees of membership or possibility indicate to what extent a datum belongs to the clusters. Possibilistic cluster analysis drops the probabilistic constraint that the sum of memberships of each datum to all clusters is equal to one.

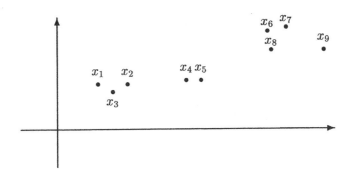

Figure 1.4: A set of data that has to be clustered

- *hierarchical cluster analysis techniques:* These techniques divide the data in several steps into more and more fine-grained classes, or they reversely combine small classes stepwise to more coarse-grained ones. Figure 1.5 shows a possible result of a hierarchical cluster analysis of the data set from figure 1.4. The small clusters on the lower levels are stepwise combined to the larger ones on the higher levels. The dashed line indicates the level in figure 1.5 that is associated with the cluster partition given in the picture.

- *objective function based cluster analysis techniques:* While hierarchical cluster analysis techniques are in general defined procedurally, i.e. by rules that say when clusters should be combined or split, the basis for the objective function methods is an objective or evaluation function that assigns each possible cluster partition a quality or error value that has to be optimized. The ideal solution is the cluster partition that obtains the best evaluation. In this sense, there is an optimization problem to be solved when applying objective function methods for cluster analysis.

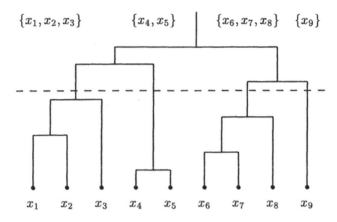

Figure 1.5: Hierarchical cluster analysis

- *cluster estimation techniques:* These techniques adopt the alternating optimization algorithm used by most objective function methods, but use heuristic equations to build partitions and estimate cluster parameters. Since the used cluster generation rules are chosen heuristically, this approach can be useful when cluster models become too complex to minimize them analytically or the objective function lacks differentiability.

The clustering techniques in chapters 2 – 5 belong to objective function methods. Cluster estimation techniques are described in chapter 6. The remaining part of this chapter is devoted to a general formal framework for fuzzy cluster analysis on the basis of objective functions.

1.3 Objective function-based cluster analysis

Before we consider fuzzy cluster analysis in more detail, we first clarify the concepts such as data space, result of a data analysis etc. that are important in the context of data and cluster analysis.

In the introductory example at the beginning of this chapter, the data space D could be the totality of all possible advertising strategies together with the predicted sales figures and production costs per piece, for instance $D := S \times \mathbb{R} \times \mathbb{R}$, if S is the set of all possible advertising strategies. A triple $(s, v_s, k_s) \in D$ then assigns the production costs $\$k_s$ per piece and a predicted sale of v_s pieces to the application of an advertising strategy s. We are interested in the advertising strategy that should be used and define the result space as $R := \{\{s\} \mid s \in D\}$. The result of the data analysis is an assignment of the given sales figures/production costs $X \subseteq D$ to the optimal advertising strategy $s \in S$. The assignment can be written as a mapping $f : X \to \{s\}$. (In this example, the pair (X, s) would also be a suitable representation of the analysis. In later examples, however, we will see the advantage of a functional definition.)

Thus, the answer to a question generally corresponds to an assignment of concretely given data $X \subseteq D$ to an (a priori unknown) answer $K \in R$ or to a mapping $X \to K$. The result space is often infinitely large so that the number of possible assignments is infinite even with a static set of data $X \subseteq D$. Each element of the analysis space is potentially a solution for a certain problem. This leads us to

Definition 1.1 (Analysis space) *Let $D \neq \emptyset$ be a set and R a set of sets with $(|R| \geq 2) \vee (\exists r \in R : |r| \geq 2)$. We call D a data space and R a result space. Then, $A(D, R) := \{f \mid f : X \to K, X \subseteq D, X \neq \emptyset, K \in R\}$ is called an analysis space. A mapping $f : X \to K \in A(D, R)$ represents the result of a data analysis by the mapping of a special, given set of data $X \subseteq D$ to a possible result $K \in R$.*

We need an evaluation criterion in order to distinguish the correct solution(s) from the numerous possible solutions. We do not want to compare the different elements of the analysis space directly using this criterion, but introduce a measure for each element. Then, this measure indirectly allows a comparison of the solutions, among other things. We introduce the evaluation criterion in the form of an objective function:

Definition 1.2 (Objective function) *Let $A(D, R)$ be an analysis space. Then, a mapping $J : A(D, R) \to \mathbb{R}$ is called an objective function of the analysis space.*

The value $J(f)$ is understood as an error or quality measure, and we aim at minimizing, respectively maximizing J. In general, we will use the objective function in order to compare different solutions for the same problem, i.e. with the same set of data. Of course, the objective function also allows an indirect comparison of two solutions for different sets of data. However, the semantics of such a comparison has to be clarified explicitly. We will not pursue this question further in this book.

In our introductory example we were interested in a rise in profit. We can evaluate a given proposal of a solution for our problem by computing the expected profit. Presupposing the same sales price of $\$d$ for all advertising strategies, the following objective function is quite canonical:

$$J : A(D, R) \to \mathbb{R}, \quad g \mapsto v_s \cdot (d - k_s) \quad \text{where} \quad g : X \to \{s\}.$$

The specification of an objective function allows us to define the answer to a given question as the (global) maximum or minimum, zero passage or another property of the objective function. The question is thus formalized with the help of an objective function and a criterion κ. The solution is defined by an element of the analysis space that fulfils κ.

Definition 1.3 (Analysis function) *Let $A(D, R)$ be an analysis space, $\kappa : A(D, R) \to \mathbb{B}$, where $\mathbb{B}$ denotes the set of the Boolean truth values, i.e. $\mathbb{B} = \{\text{true}, \text{false}\}$. A mapping $\mathcal{A} : \mathcal{P}(D) \to A(D, R)$ is called an analysis function with respect to κ if for all $X \subseteq D$:*

(i) $\mathcal{A}(X) : X \to K$, $K \in R$ and (ii) $\kappa(\mathcal{A}(X)) = \text{true}$.

For a given $X \subseteq D$, $\mathcal{A}(X)$ is called an analysis result.

In our advertisement example, κ would be defined for an $f : X \to K$ by

$$\kappa(f) = \left\{ \begin{array}{lll} true & : & J(f) = \max\{J(g)|g : X \to K \in A(D, E)\} \\ false & : & otherwise \end{array} \right.$$

Thus, when $\kappa(f)$ is valid for an $f \in A(D, R)$, f will be evaluated by J in the same way as the best solution of the analysis space. Therefore, it is the desired solution.

We still have to answer the question of how to gain the result of the analysis. Humans analyse data, for example, by merely *looking* at it. In many cases, this method is sufficient to answer simple questions. The computer analysis is carried out by an algorithm that mostly presupposes a model or a structure within the data, which the program is looking for. Here, the considered result space, the data space, and the model always have to be examined to see whether they are well-suited for the problem. The result space may be too coarse-grained or too fine-grained with respect to the considered model, or the data space may contain too little information to allow conclusions for the result space. Model design, adjustment and program development form a type of a structure analysis that is necessary for the analysis of concrete data using a computer program. Once the structure analysis has been completed, i.e. the actual question has been formalized by a specification of the spaces, the objective function and the criterion, we can look for a solution for concrete data sets. We call this process data analysis.

Definition 1.4 (Data analysis) *Let $A(D, R)$ be an analysis space, κ : $A(D, R) \rightarrow \mathbb{B}$. Then, the process of a (possibly partial) determination of the analysis function $\mathcal{A}$ with respect to κ is called data analysis.*

In our example, the data analysis consists of the determination of the element from X that provides the largest value for the objective function. Here, the process of the data analysis is well defined. Let us now investigate some further examples from the area of image processing.

Example 1   We consider black-and-white pictures of 20 by 20 points, i.e. $D := \mathbb{N}_{\leq 20} \times \mathbb{N}_{\leq 20}$, where $\mathbb{N}_{\leq t}$ denotes the set of the natural numbers without the number zero that is smaller or equal to t. We represent a picture by the set $X \subseteq D$ of the white pixels in the picture. We are interested in the brightness of the picture, measured on a scale from 0 to 1, i.e. $R := \{ \{b\} \mid b \in [0, 1]\}$. For example, a meaningful objective function would be:

$$J : A(D, R) \rightarrow \mathbb{R}, \quad f \mapsto \frac{|X|}{400} - b \quad \text{with} \quad f : X \rightarrow \{b\}.$$

The analysis of a set of data $X \subseteq D$ now consists of finding the brightness b, $\{b\} \in E$ so that $J(f) = 0$ with $f \equiv b$. In that case, the solution is obvious because of the stated J, our analysis function is

$$\mathcal{A} : \mathcal{P}(D) \rightarrow A(D, R), \quad X \mapsto f, \quad \text{with} \quad f : X \rightarrow \left\{ \frac{|X|}{400} \right\}.$$

Example 2   Let D be the same as in the previous example. We are now interested in the position of a point p with the smallest distance to all other white points in the picture, i.e. $R := \{ \{p\} \mid p \in \mathbb{N}_{\leq 20} \times \mathbb{N}_{\leq 20} \}$. As the requirement for the smallest distance, we define

$$J : A(D, R) \rightarrow \mathbb{R}, \quad f \mapsto \sum_{x \in X} ||x - p|| \quad \text{with} \quad f : X \rightarrow \{p\}.$$

The point p, $\{p\} \in R$, has the smallest distance to all other points $x \in X$ if and only if the objective function J reaches a (global) minimum for $f \equiv p$. This can be directly transformed into the analysis function:

$$\mathcal{A} : \mathcal{P}(D) \rightarrow A(D, E), \quad X \mapsto f$$

$$\text{with} \quad J(f) = \min\{J(g) \mid g : X \rightarrow K \in A(D, R)\}.$$

With a program that investigates all the 400 possible positions for p, $\mathcal{A}$ could be implemented in a naive way.

Example 3   Let D be the same as in the previous example. Each pixel represents an object. These objects have to be moved into two boxes at the positions p_1 and p_2, respectively. The question for the analysis is, which object should be moved to which box, where the distances should be as short as possible.

As the result space we choose $R := \{\{1,2\}\}$. Then, an element of the analysis space is a mapping $f : X \rightarrow \{1,2\}$. The objects that are mapped to 1 by f, shall be put into the first box, the other objects into the second. The objective function is of course the sum of all distances to be covered:

$$J : A(D,R) \rightarrow \mathbb{R}, \quad f \mapsto \sum_{x \in X} \|x - p_{f(x)}\| \quad \text{with} \quad f : X \rightarrow \{1,2\}.$$

Again, the minimum of the objective function has to be found. The sum of the distances can be minimized by minimizing the individual distances. So, if $c_x \in \{1,2\}$ is the closer box for the object $x \in X$, the analysis function results in:

$$\mathcal{A}(X) = f \quad \text{with} \quad f : X \rightarrow \{1,2\}, \quad x \mapsto c_x.$$

The previous example is of special interest. For the first time, the information gained from the analysis was not related to the complete set of data but individually to each element of the data set. That was possible because the elements of the result space had multiple elements themselves. Thus, the result of the analysis is a partition of the data, a class or cluster partition. (Here, the advantage of the functional definition becomes apparent as already mentioned on page 12.)

Definition 1.5 (Cluster partition) *Let $A(D,R)$ be an analysis space, $X \subseteq D$, $f : X \rightarrow K \in A(D,R)$, $A_k := f^{-1}(k)$ for $k \in K$. Then, f is called a cluster partition if $\{A_k \mid k \in K\}$ is a (hard) partition of X, i.e.*

$$\bigcup_{i \in K} A_i = X \tag{1.1}$$

$$\forall i,j \in K : \quad i \neq j \Rightarrow A_i \cap A_j = \emptyset \tag{1.2}$$

$$\forall i \in K : \quad \emptyset \neq A_i \neq X. \tag{1.3}$$

Remark 1.6 (Cluster partition) *$f : X \rightarrow K$ is a cluster partition if and only if f is exhaustive and $|X|, |K| \geq 2$.*

Proof: $\Leftarrow$: Let f be exhaustive, $|X|, |K| \geq 2$. We have to show (1.1) to (1.3). (1.1) : Let $x \in X$. This results in $x \in f^{-1}(f(x)) = A_{f(x)} \subseteq \bigcup_{k \in K} A_k$. $A_i \subseteq X$, $i \in K$, is obviously valid. (1.2) : Let $i, j \in K$ with $i \neq j$. Let $x \in A_i$, i.e. $f(x) = i$. Since f is unambiguous as a mapping, it follows that $f(x) \neq j$, i.e. $x \notin A_j$. (1.3) : Let $i \in K$. Since f is exhaustive, one can find an $x \in X$ with $f(x) = i$. It follows that $x \in A_i$ and thus $A_i \neq \emptyset$. Because of $|K| \geq 2$, there exists a $j \in K$ with $i \neq j$. Also, there must be a $y \in X$ with $y \in A_j$. From (1.2) follows $y \notin A_i$, i.e. $A_i \neq X$.
$\Rightarrow$: Let $f : X \to K$ be a cluster partition. (1.1) to (1.3) hold. Since f is an analysis result, $X \neq \emptyset$ follows corresponding to the definition. From (1.1) also follows that $K \neq \emptyset$. Would X have exactly one element, there would be an $i \in K$ with $A_i = X$ because of (1.1), which contradicts (1.3). It follows that $|X| \geq 2$. From (1.3) also follows that $|K| \geq 2$. Let $i \in K$. Because of $A_i \neq \emptyset$ from (1.3), one can find an $x \in A_i \subseteq X$ so that $f(x) = i$ follows, i.e. f is exhaustive. ■

The possibilities for the data analysis are of course not exhausted by simple result spaces like $R = \{\{1, 2\}\}$ from the previous example.

Example 4   Let the data space D be the same as in the previous example. Circles are to be found in the picture and the data are to be assigned to the circles. If we know that the picture contains c circles we can use $\{\{1, 2, ..., c\}\}$ as the result space. Thus, we obtain a cluster partition for the numbers 1 to c, each of which represents a circle. If we are interested in the exact shape of the circles, we can characterize them by a triple $(x, y, r) \in \mathbb{R}^3$, where x, y stand for the coordinates of the centre and $|r|$ for the radius. If we set $R := \mathcal{P}_c(\mathbb{R}^3)$, the analysis result is $f : X \to K$, $K \in R$, a cluster partition of the data to c circles. For a $z \in X$, $f(z) = (x, y, r)$ is the assigned circle.
If the number of circles is not known we can choose in the first case $R := \{\{1\}, \{1, 2\}, \{1, 2, 3\}, ...\} = \{\mathbb{N}_{\leq k} \mid k \in \mathbb{N}\}$, and in the second case $R := \{K \mid K \subseteq \mathbb{R}^3, K \text{ finite}\}$. The result of the analysis $f : X \to K$ can be interpreted as previously, but now, additionally, the number of those circles that were recognized is provided by $|K|$.

Moreover, as a modification of example 3 we could be interested in giving both boxes suitable positions so that the distances between the data points and the boxes become as short as possible. (For instance, this kind of analysis is carried out by the algorithm in section 2.1.)

1.4 Fuzzy analysis of data

A deterministic cluster partition as the result of a data analysis is equivalent to a hard partition of the data set. Although this formalization corresponds to what we had in mind in the previous examples, it turns out to be unsuitable on closer inspection. Let us imagine that we are looking for two overlapping circles in the picture. Further, let us assume that there are pixels exactly at the intersection points of the two circles. Then, a cluster partition assigns these pixels to exactly one circle, forced by the characteristics of a cluster partition. This is also the case for all other data that have the same distance from the two contours of the circles (figure 1.6). However, it can not be accepted that these pixels are assigned to one or the other circle. They *equally* belong to both circles. Here, we have one of these cases already mentioned, where the choice of the result space is not adequate for the considered question. The result space is too *coarse-grained* for a satisfying assignment of the data.

A solution for this problem is provided by the introduction of gradual memberships to fuzzy sets [104]:

Definition 1.7 (Fuzzy set) *A fuzzy set of a set X is a mapping $\mu : X \to [0,1]$. The set of all fuzzy sets of X is denoted by $F(X) := \{\mu | \mu : X \to [0,1]\}$.*

For a fuzzy set μ_M, there is – besides the *hard* cases $x \in M$ and $x \notin M$ – a smooth transition for the membership of x to M. A value close to 1 for $\mu_M(x)$ means a high degree of membership, a value close to 0 means a low degree of membership. Each conventional set M can be transformed into a fuzzy set μ_M by defining $\mu_M(x) = 1 \Leftrightarrow x \in M$ and $\mu_M(x) = 0 \Leftrightarrow x \notin M$.

Since we want to fuzzify the membership of the data to the clusters in our examples, it seems convenient to consider now $g : X \to F(K)$ instead of $f : X \to K$ as the result of the analysis. A corresponding interpretation for this new result of the analysis would be that $g(x)(k) = 1$, if x can unambiguously be assigned to the cluster k, and $g(x)(k) = 0$, if x does definitely not belong to the cluster k. A gradual membership such as $g(x)(k) = \frac{1}{3}$ means that the datum x is assigned to the cluster k with the degree of one third. Statements about memberships to the other clusters are made by the values of $g(x)(j)$ with $k \neq j$. Thus, it is possible to assign the datum x to the clusters i and j in equal shares, by setting $g(x)(i) = \frac{1}{2} = g(x)(j)$ and $g(x)(k) = 0$ for all $k \in K\backslash\{i,j\}$. This leads to the following definitions of an analysis space and a cluster partition, generalized to fuzzy sets:

Definition 1.8 (Fuzzy analysis space) *Let* $A(D, R)$ *be an analysis space. Then,* $A_{\text{fuzzy}}(D, R) := A(D, \{F(K)|K \in R\})$ *defines a further analysis space, the fuzzy analysis space for* $A(D, R)$. *The results of an analysis are then in the form* $f : X \to F(K)$ *for* $X \subseteq D$ *and* $K \in R$.

Definition 1.9 (Probabilistic cluster partition) *Let* $A_{\text{fuzzy}}(D, R)$ *be an analysis space. Then, a mapping* $f : X \to F(K) \in A_{\text{fuzzy}}(D, R)$ *is called a probabilistic cluster partition if*

$$\forall x \in X : \quad \sum_{k \in K} f(x)(k) = 1 \quad \text{and} \tag{1.4}$$

$$\forall k \in K : \quad \sum_{x \in X} f(x)(k) > 0 \tag{1.5}$$

hold. We interpret $f(x)(k)$ *as the degree of membership of the datum* $x \in X$ *to the cluster* $k \in K$ *relative to all other clusters.*

Although this definition strongly differs from the definition of a cluster partition at first glance, the differences are rather small; they only *soften* the conditions (1.1) to (1.3). The requirement (1.2) for disjoint clusters has to be modified accordingly for gradual memberships. The condition (1.4) means that the sum of the memberships of each datum to the clusters is 1, which corresponds to a normalization of the memberships per datum. This means that each individual datum receives the same weight in comparison to all other data. This requirement is also related to condition (1.1) because both statements express that all data are (equally) included into the cluster partition. The condition (1.5) says that no cluster k can be empty, i.e. the membership $f(x)(k)$ must not be zero for all x. This corresponds to the inequality $A_i \neq \emptyset$ from (1.3). By analogy with the conclusion in remark 1.6, it follows that no cluster can obtain all memberships ($A_i \neq X$ in (1.3)).

The name *probabilistic* cluster partition refers to an interpretation in the sense of *probabilities*. It suggests an interpretation like "$f(x)(k)$ *is the probability for the membership of x to a cluster k*". However, this formulation is misleading. One can easily confuse the degree in which a datum x represents a cluster k with the probability of an assignment for a datum x to a cluster k.

Figure 1.6 shows two circles with some pixels that are equidistant from both circles. The two data items close to the intersection points of the circles can be accepted as typical representatives of the circle lines. This does not apply to the other data with their increasing distances to the circles. Intuitively, the membership of the more distant points to the circles should be very low. Because of the normalization, however, the sum of the memberships has to be 1, so the individual memberships have to be approximately $\frac{1}{2}$. Being unaware of the normalization, the fairly high values for

| Figure 1.6: Data that have unambiguously to be assigned to one of the circles in a non-deterministic way when applying a hard partition. | Figure 1.7: Growing degrees of membership to a cluster do not necessarily mean better representatives for the cluster. |

the membership could (without the clarifying figure) leave the impression that the data are rather typical for the circles after all. A similar case is shown in figure 1.7. This time, all data are very far away from both circles, but the distances of the data on the left-hand side and on the right-hand side from both circles are different. The datum in the middle will again receive memberships of around $\frac{1}{2}$, while the outliers are assigned to the respective closer circle by memberships clearly above $\frac{1}{2}$. All data are (almost) equally untypical for the circles, but the memberships are very different. In order to avoid misinterpretations, a better reading of the memberships would be *"If x has to be assigned to a cluster, then with the probability $f(x)(k)$ to the cluster k"*.

This semantics has emerged from the definition only, thus it is not at all cogent. Just as well, as a semantics could be forced (per definition) that would in fact allow us to conclude from high degrees of membership for a datum x to a cluster k that x is a typical representative for k. In this case we should drop the condition (1.4) which is responsible the normalization of the memberships. Then, the more distant data in the figures 1.6 and 1.7 could receive lower degrees of membership than $\frac{1}{2}$.

Definition 1.10 (Possibilistic cluster partition) *Let $A_{\text{fuzzy}}(D, R)$ be an analysis space. Then, a result of an analysis $f : X \rightarrow F(K)$ is called possibilistic cluster partition if*

$$\forall k \in K : \quad \sum_{x \in X} f(x)(k) > 0 \tag{1.6}$$

holds. We interpret $f(x)(k)$ as the degree of representativity or typicality of the datum $x \in X$ for the cluster $k \in K$

Possibilistic cluster partitions are useful especially in the analysis of images, because there are often disturbances that actually can not be assigned to one of the clusters. Mostly, those disturbances are responsible for non-satisfying analysis results of a probabilistic cluster analysis; we will see examples of that later.

As far as the formal part of the definition is concerned, every probabilistic cluster partition is also a possibilistic partition. However, since a probabilistic partition obtains a completely different meaning with the possibilistic interpretation, we have included the semantics of the partition in the definition.

We have to emphasize that the type of fuzzification that we have carried out here is not at all the only possibility. In our approach, the analysis itself was fuzzified, meaning that the result of the analysis, i.e. the partition of the data, is fuzzy. Thus, this section is called *fuzzy analysis of data*. In a similar way, the data space could have been fuzzified; fuzzy image data would have been allowed in our example, for example, grey values. However, we will not pursue the topic of *analysis of fuzzy data* in this book, since we usually have to deal with data in the form of crisp measurements.

1.5   Special objective functions

Returning to the probabilistic cluster partitions, let us consider the following

Example 5   A picture is showing the contours of the tyres of a bicycle. For simplicity, we assume that the positions of the front axle and the rear axle are known. The picture will be represented as $X \subseteq \mathbb{R}^2$, similar to the previous examples. A pixel is set if its pair of coordinates is an element of X. We are interested in the radii of the wheels (besides the probabilistic cluster partition).

For this example the following definitions might be suitable. $D := \mathbb{R}^2$, $W := \{ (k,r) \mid k \in \{\text{front wheel}, \text{rear wheel}\}, r \in \mathbb{R}\}$, $R := \{ \{(\text{front wheel}, r), (\text{rear wheel}, s)\} \mid r, s \in \mathbb{R}\} \subseteq \mathcal{P}_2(W)$. We denote the set of subsets of the set M that have t elements by $\mathcal{P}_t(M)$. As an analysis space, we use $A_{fuzzy}(D, R)$. Let the position of the axles be defined as $p_{\text{frontwheel}}, p_{\text{rearwheel}} \in \mathbb{R}^2$. Now, we have to provide an objective function. A large family of objective functions results from the following basic function (generalization of the least squared error):

$$J(f) = \sum_{x \in X} \sum_{k \in K} f^m(x)(k) \cdot d^2(x,k) \quad \text{with} \quad f : X \to F(K). \qquad (1.7)$$

Here, $d(x, k)$ is a measure for the distance between the datum x and the cluster k. There is no occurrence of $f(x)$ in this *distance function*. Thus, the distance does not depend on the memberships. We use square distances so that the objective function is non-negative. The exponent $m \in \mathbb{R}_{>1}$ – the so-called *fuzzifier* – represents a weight parameter. We denote the set of all real numbers greater than t by $\mathbb{R}_{>t}$. This way, the search for an objective function has been simplified, because the choice of a real-valued parameter m and a distance function d is certainly simpler than the choice of an arbitrary function $J : A(D, R) \to \mathbb{R}$.

As a distance function, we choose in our example

$$d : \mathbb{R}^2 \times R \to \mathbb{R}, \quad (x, (k, r)) \mapsto \left| \|x - p_k\| - r \right|.$$

What is the effect of this objective function? Let us look at a pixel $x \in X$ that is located close to the front wheel. The distance function $d(x, k)$ yields the value 0 if and only if x is on the wheel k. For $d(x, (\text{front wheel}, r_v))$, we thus obtain a value near 0. For $d(x, (\text{rear wheel}, r_h))$, however, we obtain a greater value, since the distance to the rear wheel is clearly greater. Let $f \in A_{\text{fuzzy}}(D, R)$ be a probabilistic cluster partition of X, where the data are assigned to the corresponding wheels. Then, $f(x)(\text{front wheel}, r_v) \approx 1$ and $f(x)(\text{rear wheel}, r_h) \approx 0$. Because of the product $f^m(x)(k) \cdot d^2(x, k)$, the sum is small in both cases, since in the first case the distance to the front wheel is small and in the second case the membership to the rear wheel is low. The intersection of the two wheels is impossible in our example, however, it would not falsify our calculation because the memberships to both wheels would be approximately $\frac{1}{2}$ for the intersection points, while the distances would be small in both cases.

Thus, the criterion for the objective function is also obvious: it has to be minimized at the solution. In practice, (1.7) is sufficient for the solution of many problems. Depending on the problem the distance measure has to be modified.

Let us now take a closer look at the influence of the *fuzzifier* m. The greater m is the faster $f^m(x)(k)$ will become 0, e.g. with $m = 6$, relatively high memberships such as 0.8 are decreased to a factor of about 0.26. Thus, (local) minima and maxima will be less clearly developed or might even completely vanish. The greater m, the *fuzzier* the results will be. We can choose m depending on the estimation of how well the data can be divided into clusters. For instance, if the clusters are far away from each other, even a crisp partition is possible. In this case, we choose m close to 1. (If m gets close to 1, the memberships converge at 0 or 1, cf. [10].) If the clusters are hardly distinguishable, m should be chosen very large. (If m goes to infinity, the memberships become to $\frac{1}{c}$ where c is the number of clusters.)

Of course, some experience is necessary for this decision. Moreover, a data set can contain clusters with different separabilities. A common choice of the fuzzifier is $m = 2$.

In any case, the (global) minimum of the objective function with the constraint of a probabilistic cluster partition is the aim of the data analysis. However, this is often not as simple as in the first examples in this chapter. An exhaustive search is not feasible because of the large number of possibilities. Moreover, the objective function is multi-dimensional, so that the well-known one-dimensional techniques to search for minima can not be applied. Therefore, we will have to be satisfied with local minima from the start.

How do we choose the memberships $f(x)(k)$, $x \in X$, $k \in K$ in order to obtain a minimum of the objective function with an a priori known cluster set K? The answer is given in the following theorem [10].

Theorem 1.11 *Let $A_{\text{fuzzy}}(D, R)$ be an analysis space, $X \subseteq D$, $C \in R$, J an evaluation function corresponding to (1.7), d the corresponding distance function and $m \in \mathbb{R}_{>1}$. If the objective function has a minimum for all probabilistic cluster partitions $X \to F(K)$ at $f : X \to F(K)$ then*

$$
f(x)(k) \mapsto
\begin{cases}
\dfrac{1}{\sum_{j \in K} \left(\frac{d^2(x,k)}{d^2(x,j)} \right)^{\frac{1}{m-1}}} & : \quad \text{for } I_x = \emptyset \\[2ex]
\sum_{i \in I_x} f(x)(i) = 1 & : \quad \text{for } I_x \neq \emptyset, \ k \in I_x \\[2ex]
0 & : \quad \text{for } I_x \neq \emptyset, \ k \notin I_x,
\end{cases}
$$

where $I_x := \{ j \in K \,|\, d(x,j) = 0 \}$.

If a set I_x contains more than one element, $f(x)(k)$ for $k \in I_x$ is not uniquely determined. We will discuss this problem on page 24.

Proof: A necessary criterion for a minimum is a zero passage in the first derivative. Because of the high dimensionality of the objective function, we will restrict ourselves here to zero passages of partial derivatives.

Let $f : X \to F(K)$ be a probabilistic cluster partition, $x \in X$, define $u := f(x)$, $u_k := f(x)(k)$ for $k \in K$. Let the objective function J be minimal for all probabilistic cluster partitions in f. Thus, the conditions (1.4) and (1.5) hold for f. We can minimize $\sum_{x \in X} \sum_{k \in K} f^m(x)(k) \cdot d^2(x,k)$ by minimizing the sum $\sum_{k \in K} u_k^m \cdot d^2(x,k)$ for all $x \in X$. The validity of the constraint (1.4) can be expressed by introducing a Lagrange multiplier λ also by a zero passage in the partial derivative with respect to λ. Therefore, we define for $x \in X$:

$$J_x := \sum_{k \in K} u_k^m d^2(x, k) - \lambda \left(\left(\sum_{k \in K} u_k \right) - 1 \right).$$

For all $k \in K$, the partial derivatives

$$\frac{\partial}{\partial \lambda} J_x(\lambda, u) = \left(\sum_{k \in K} u_k \right) - 1$$

$$\frac{\partial}{\partial u_k} J_x(\lambda, u) = m \cdot u_k^{m-1} \cdot d^2(x, k) - \lambda$$

must be zero. We use use the fact that the distance function is independent of the memberships. Let us first consider the case that the distances are non-zero. In this case, we derive from the previous equation for each $k \in K$:

$$u_k = \left(\frac{\lambda}{m \cdot d^2(x, k)} \right)^{\frac{1}{m-1}}. \tag{1.8}$$

Using this fact in $\frac{\partial}{\partial \lambda} J_x(\lambda, u) = 0$ leads to

$$1 = \sum_{j \in K} u_j$$

$$= \sum_{j \in K} \left(\frac{\lambda}{m \cdot d^2(x, j)} \right)^{\frac{1}{m-1}}$$

$$= \left(\frac{\lambda}{m} \right)^{\frac{1}{m-1}} \sum_{j \in K} \left(\frac{1}{d^2(x, j)} \right)^{\frac{1}{m-1}}.$$

Thus,

$$\left(\frac{\lambda}{m} \right)^{\frac{1}{m-1}} = \frac{1}{\sum_{j \in K} \left(\frac{1}{d^2(x, j)} \right)^{\frac{1}{m-1}}}.$$

Together with (1.8), we obtain for the membership

$$u_k = \frac{1}{\sum_{j \in K} \left(\frac{1}{d^2(x, j)} \right)^{\frac{1}{m-1}}} \cdot \left(\frac{1}{d^2(x, k)} \right)^{\frac{1}{m-1}}$$

$$= \frac{1}{\sum_{j \in K} \left(\frac{d^2(x, k)}{d^2(x, j)} \right)^{\frac{1}{m-1}}}.$$

A special case, that we have to consider separately, is $d(x,j) = 0$ for at least one $j \in K$. In this case, the datum x is lying exactly in cluster j. Therefore, the set $I_x := \{k \in K \mid d(x,k) = 0\}$ contains all clusters where the datum x *suits exactly*. Here, the minimization problem can be solved easily because $J_x(\lambda, u) = 0$ and (1.4) hold if $u_k = 0$ for all $k \notin I_x$ and $\sum_{j \in I} u_j = 1$. ∎

In the implementation of an algorithm to search for a probabilistic cluster partition, where we of course do not start from a minimum but want to find it, we have to take care of condition (1.5). This condition must hold, because the mapping $f : X \to F(K)$ has to be a cluster partition. However, it does not lead to any greater problems when (1.5) does not hold. If there really exists a $k \in K$ with $\sum_{x \in X} f(x)(k) = 0$, we actually have a cluster partition into the cluster set $K \setminus \{k\}$. We can remove all superfluous clusters, and obtain a correct cluster partition for which (1.5) hold.

Under certain conditions, we can be sure that $\sum_{x \in X} f(x)(k) = 0$ can not occur. From the theorem we can see that condition (1.5) automatically holds, when there is at least one $x \in X \subseteq D$ with $f(x)(k) \neq 0$ for each $k \in K$. The membership $\sum_{k \in K} \frac{d^2(x,k)}{d^2(x,j)}$ is then a positive value for each $k \in K$. If we require for the set of data that there always exists a datum that has a positive distance to all clusters, (1.5) is always fulfilled.

If a single cluster is characterized by point (the centre of the cluster), a metric can be used as the distance function. Then the distance to a cluster can vanish for at most *one* datum, i.e. if there are more data than clusters, condition (1.5) always holds. For its validity, the prerequisite $|X| > |K|$ is sufficient. On the other hand, if contours of circles describe the cluster shapes, there can potentially be infinitely many points that are located on the circle, i.e. they have the distance 0. Since a circle is uniquely determined by three points, perhaps the following condition could be formulated: "$|X| > 3|K|$ *and there exists an $x \in X$ with a positive distance to all circles for all possible partitions*". However, a condition like that cannot be proved in practice. If it could be done, the question of the cluster partition could be answered easily. Therefore, in practice we proceed as mentioned above: if condition (1.5) does not hold, we remove a cluster and restart the analysis.

Regarding the implementation, the theorem is not specific in the case $|I_x| > 1$. In any case, the sum of memberships has to be 1. For the implementation, concrete values have to be provided for each u_j with $j \in I_x$.

In the example with two overlapping circles, where data are located on the intersection points of the circles, a membership $u_j := \frac{1}{|I_x|}$ corresponds best to the chosen semantics. However, if we deal with point clusters (with a metric as distance function), $|I_x| > 1$ means that several clusters are identical. In such a case, it is better to choose different membership degrees to the identical clusters such that the clusters get separated again in the next step. However, if the clusters remain indistinguishable, it is again basically a data analysis with a cluster set decreased by one element.

The question from theorem 1.11 still has to answered for the case of a possibilistic cluster partition. Here, we can expect a simpler expression, since the constraint (1.4) can be dropped. However, we get a trivial solution without this constraint, viz. we simply choose $f(x)(k) = 0$ for $f : X \to F(K)$ and all $x \in X$, $k \in K$ and obtain a global minimum of the objective function. Thus, we can not simply reuse the objective function (1.7). We define instead, following a proposal of Krishnapuram and Keller [62]:

$$J : A(D, R) \to \mathbb{R},$$

$$f \mapsto \sum_{x \in X} \sum_{k \in K} f^m(x)(k) \cdot d^2(x, k) + \sum_{k \in K} \eta_k \sum_{x \in X} (1 - f(x)(k))^m. \qquad (1.9)$$

The first sum is taken directly from the objective function (1.7). The second sum *rewards high memberships*, since memberships $f(x)(k)$ close to 1 make the expression $(1 - f(x)(k))^m$ become approximately 0. Here, it should be stressed again that now *each* membership $f(x)(k)$ can be almost 1 (possibilistic clustering, see the example with the intersection of two circles). Because the high memberships also imply a reliable assignment of the data to the clusters, we achieve the desired effect. Before we discuss the additional factors η_k for $k \in K$, let us formulate the following theorem:

Theorem 1.12 Let $A_{\text{fuzzy}}(D, R)$ be an analysis space, $X \subseteq D$, $K \in R$, J an objective function corresponding to (1.9), d the corresponding distance function and $m \in \mathbb{R}_{>1}$. If the objective function reaches its minimum for all possibilistic cluster partitions $X \to F(K)$ at $f : X \to F(K)$ then

$$f(x)(k) \mapsto \frac{1}{1 + \left(\frac{d^2(x,k)}{\eta_k} \right)^{\frac{1}{m-1}}},$$

holds, where $\eta_k \in \mathbb{R}$ for $k \in K$.

Proof: Let $u_{k,x} := f(x)(k)$. Like in the previous theorem, we compute all partial derivatives $\frac{\partial}{\partial u_{k,x}} J$. They have to vanish for a minimum:

$$
\begin{aligned}
0 &= \frac{\partial}{\partial u_{k,x}} J \\
&= m \cdot d^2(x,k) \cdot u_{k,x}^{m-1} - \eta_k \cdot m(1 - u_{k,x})^{m-1} \\
\Leftrightarrow \quad u_{k,x}^{m-1} \cdot d^2(x,k) &= \eta_k (1 - u_{k,x})^{m-1} \\
\Leftrightarrow \quad \frac{d^2(x,k)}{\eta_k} &= \left(\frac{1 - u_{k,x}}{u_{k,x}}\right)^{m-1} = \left(\frac{1}{u_{k,x}} - 1\right)^{m-1} \\
\Leftrightarrow \quad u_{k,x} &= \frac{1}{1 + \left(\frac{d^2(x,k)}{\eta_k}\right)^{\frac{1}{m-1}}} .
\end{aligned}
$$

■

Let us now explain the factors η_k, $k \in K$. We consider the case $m = 2$ and the requirement of the theorem $u_{k,x} = (1 + \frac{d^2(x,k)}{\eta_k})^{-1}$. If $\eta_k = d^2(x,k)$ is valid, then $u_{k,x} = (1 + \frac{d^2(x,k)}{d^2(x,k)})^{-1} = (1 + 1)^{-1} = \frac{1}{2}$. With the parameter η_k, we thus fix for each cluster $k \in K$, for which distance the membership should be $\frac{1}{2}$. If we consider $\frac{1}{2}$ as the limit for a definite assignment of a datum x to cluster k, the meaning of η_k is obvious. We control the permitted expansion of the cluster with this parameter. If we consider circle-shaped full clusters, $\sqrt{\eta_k}$ approximately corresponds to the mean diameter of the cluster; if we consider the rim of the circles' as clusters, $\sqrt{\eta_k}$ corresponds to the mean thickness of the contour. If the shapes, that are to be determined by the analysis, are known in advance, η_k can easily be estimated. If all clusters have the same shape, the same value can be chosen for all clusters. However, this kind of a priori knowledge is generally not available. In this case we can estimate

$$
\eta_k = \frac{\sum_{x \in X} f^m(x)(k) d^2(x,k)}{\sum_{x \in X} f^m(x)(k)} \tag{1.10}
$$

for each $k \in K$. Here, the mean distance is chosen. In the literature [62], one can find further suggestions for the choice or estimation of η_k.

The objective function (1.7) with the probabilistic constraint definitely aims at partitioning the data set, at least in a fuzzy manner. The objective function (1.9) treats each cluster independently. Therefore, one cluster in the data set can be shared by two clusters in the possibilistic cluster partition without affecting the value of the objective function. In this sense, the number c specifies the desired number of clusters in the probabilistic case, whereas it represents only an upper bound in the possibilistic case. This effect was also observed in [7] that possibilistic clustering can produce identical clusters and at the same time interpret other clusters contained in the data as noise. Krishnapuram and Keller [63] emphasize that probabilistic clustering is primarily a partitioning algorithm, whereas possibilistic clustering is a rather mode-seeking technique, aimed at finding meaningful clusters.

An alternative approach called *noise clustering* was proposed by Davé [25]. An early attempt to take noise into account was described by Ohashi [80] (see also [28]). The objective function (1.7) was modified to

$$J(f) = \alpha \sum_{x \in X} \sum_{k \in K} f^m(x)(k) \cdot d^2(x,k) + (1-\alpha) \sum_{x \in X} f^m(x)(k_{\text{noise}}) \quad (1.11)$$

where $f : X \to F(K \cup \{k_{\text{noise}}\})$ so that a noise cluster k_{noise} is added to the set of clusters K. Of course, the probabilistic constraints from definition 1.9 are in this case

$$\forall x \in X : \qquad \sum_{k \in K \cup \{k_{\text{noise}}\}} f(x)(k) = 1 \quad \text{and} \qquad (1.12)$$

$$\forall k \in K : \qquad \sum_{x \in X} f(x)(k) > 0. \qquad (1.13)$$

The objective function (1.11) is minimized when those data (outliers) that are far away from all clusters in K are assigned to the noise cluster k_{noise} with a high membership degree.

Independently of Ohashi, Davé [25] introduced the concept of noise clustering by a slightly modified version of the objective function (1.11) that has a very simple interpretation:

$$J(f) = \sum_{x \in X} \sum_{k \in K} f^m(x)(k) \cdot d^2(x,k)$$
$$+ \sum_{x \in X} \delta^2 \left(1 - \sum_{k \in K} f(x)(k)\right)^m \qquad (1.14)$$

The first term again coincides with the objective function for probabilistic clustering. The second term is intended to represent a noise cluster. δ has to be chosen in advance and is supposed to be the distance of all data to a noise cluster. For this objective function we can even drop the probabilistic constraint, since it is already incorporated in the term for the noise cluster. The membership degree of a datum to the noise cluster is defined as one minus the sum of the membership degrees to all other clusters. If the distances of a datum x to all clusters are about δ or even greater, then the minimization of (1.14) enforces that the values $f(x)(k)$ have to be small and the membership degree $1 - \sum_{k \in K} f(x)(k)$ to the noise cluster is larger. For the case $m = 1$ that would lead to a crisp assignment of the data to the clusters, a datum would be assigned to the noise cluster if and only if its distances to all other clusters are greater than δ.

In opposition to possibilistic clustering where the probabilistic constraint (1.4) is completely dropped, i.e. $\sum_{k \in K} f(x)(k) > 1$ and $\sum_{k \in K} f(x)(k) < 1$ are both possible, noise clustering admits only the latter inequality.

It is easily seen that the minimzation of Davé's objective function (1.14) is equivalent to the minimization of Ohashi's version (1.11) when $\alpha = \dfrac{1}{1 + \delta^2}$ is chosen.

When we optimize the membership degrees $f(x)(k)$, the optimization of the (noise) objective function (1.14) without constraints is equivalent to optimizing the objective function (1.7) with the probabilistic constraint and $c + 1$ clusters. The additional cluster corresponds to the noise cluster k_{noise} for which $d(x, k_{\text{noise}}) = \delta$ is assumed. Therefore, we directly derive from theorem 1.11 a necessary condition for the minimization of the objective function (1.14) for noise clustering:

$$f(x)(k) \mapsto \frac{1}{\sum_{j \in K} \left(\dfrac{d^2(x,k)}{d^2(x,j)} \right)^{\frac{1}{m-1}} + \left(\dfrac{d^2(x,k)}{\delta^2} \right)^{\frac{1}{m-1}}} \tag{1.15}$$

In [26] noise clustering is shown to be a generalized combination of the probabilistic scheme in theorem 1.11, and the possibilistic scheme in theorem 1.12, by allowing individual distances of the data to the noise cluster instead of the fixed distance δ for all data.

1.6 A principal algorithm for a known number of clusters

After the principal model of the data analysis is chosen, i.e. the analysis space, the objective function etc., it is desirable to use a computer for the analysis of the data. For the case, in which an objective function has one of the forms from the previous section, we now describe an algorithm that carries out the data analysis. The technique has its origin in the hard c-means algorithm (see for instance [30]), which provides a cluster partition corresponding to the definition 1.5. The first generalization to probabilistic cluster partitions came from Dunn [31]. With this algorithm, an objective function of the form (1.7) was used, where $m = 2$ was chosen. The generalization to arbitrary values $m \in \mathbb{R}_{>1}$ stems from Bezdek [8]. All variants described in [8] are tailored for the recognition of spherical clouds of points in the data set. The more general investigations on the convergence properties of the fuzzy c-means algorithm [9, 14] are not restricted only to purely spherical clouds of points [10].

The algorithm presented here provides a probabilistic cluster partition of the given data. With regard to the implementation, the result of an analysis may be in an inconvenient form, because functional programming languages mainly allow functions as data types. We provide a procedural description here for reasons of efficiency. However, if $X = \{x_1, x_2, \ldots, x_n\}$ and $K = \{k_1, k_2, \ldots, k_c\}$ are finite, an analysis result $f : X \to F(K)$ can be represented as a $c \times n$-matrix U, where $u_{i,j} := f(x_j)(k_i)$. The algorithm is searching for an analysis result $f \in A_{\text{fuzzy}}(D, R)$ that minimizes an objective function. This minimization is realized with an iteration procedure where in each step the matrix U and the cluster set $K \in R$ will be as optimal as possible and adjusted to each other (U and K completely characterize f). Bezdek, Hathaway and Pal [13] call this technique *alternating optimization* (AO).

Algorithm 1 (probabilistic clustering algorithm)

> *Let a data set $X = \{x_1, x_2, \ldots, x_n\}$ be given. Let each cluster be uniquely characterizable by an element of a set K.*
>
> *Choose the number c of clusters, $2 \le c < n$*
> *Choose an $m \in \mathbb{R}_{>1}$*
> *Choose a precision for termination ε*
> *Initialize $U^{(0)}$, $i := 0$*
> *REPEAT*
>   *Increase i by 1*
>   *Determine $K^{(i)} \in \mathcal{P}_c(K)$ such that J is minimized*
>   *by $K^{(i)}$ for (fixed) $U^{(i-1)}$*
>   *Determine $U^{(i)}$ according to theorem 1.11*
> *UNTIL $\|U^{(i-1)} - U^{(i)}\| \le \varepsilon$*

The search for the optimal set of clusters $K^{(i)} \in \mathcal{P}_c(K)$ is naturally dependent on the distance function used in the objective function. It has to be adjusted individually to the respective data analysis for this reason and will not be treated in more detail in this section. When introducing the special techniques in the chapters 2, 3, 4, and 5, the choice of the cluster set will be discussed in detail. The step towards the optimization of the memberships, however, is given by theorem 1.11 for all techniques in the same way.

Unfortunately, no general result on the convergence for all techniques that are based on algorithm 1 is known at the moment. Bezdek has shown for his algorithm in [10] that either the iteration sequence itself or any convergent subsequence converges in a saddle point or a minimum – but not in a maximum – of the objective function. The proof is based on the convergence theorem of Zangwill which has also been used to prove the convergence of numerous classical iteration techniques. For other clustering techniques convergence proofs were also provided, but there are no proofs for most of the more recent shell clustering techniques from chapter 4. However, the promising results justify their application, too.

Analogous to the probabilistic one, we define a possibilistic clustering algorithm. It seems natural to simply change the algorithm 1 so that the memberships are no longer computed according to theorem 1.11, but corresponding to theorem 1.12. However, this does not lead to satisfying results in general. The algorithm shows the tendency to interpret data with a low membership in all clusters (because of memberships close to 0) as *outliers* instead of further adjusting the possibly non-optimal cluster set K to these data. Therefore, a probabilistic data analysis is carried out before. The result of the analysis is used as an initialization for the following steps, especially for the determination of the η_k, $k \in K$. The possibilistic data analysis is carried out now with these initial values. Finally, the η_k, $k \in K$ are estimated once more and the algorithm is run again.

The determination of the set of clusters $K^{(i)}$ is done here in each iteration step in the same way as with algorithm 1. How the $K^{(i)}$ are actually computed is described in detail in the chapters where the corresponding clustering technique is introduced.

The noise clustering algorithm is identical to the probabilistic algorithm 1, except that

- the distance δ to the noise cluster has to be chosen in addition to the values c, m, and ε, and

- the matrix $U^{(i)}$ has to be determined according to equation (1.15) instead of the formula in theorem 1.12.

From a pragmatic point of view, it is in most cases easier to start the algorithms by initializing the prototypes $K^{(i)}$ instead of the membership matrix $U^{(0)}$. The termination criterion $||U^{(i-1)} - U^{(i)}|| \leq \varepsilon$ can also be modified to $||K^{(i-1)} - K^{(i)}|| \leq \varepsilon$, i.e. the algorithm terminates when the

Algorithm 2 (possibilistic clustering algorithm)

Let a data set $X = \{x_1, x_2, \ldots, x_n\}$ be given. Let each cluster be uniquely characterizable by an element of a set K.

Choose the number c of clusters, $2 \leq c < n$
Choose an $m \in \mathbb{R}_{>1}$
Choose a precision for termination ε
Execute algorithm 1
FOR 2 TIMES
    Initialize $U^{(0)}$ and $K^{(0)}$ with the previous results, $i := 0$
    Initialize η_k for $k \in K$ according to (1.10)
    REPEAT
        Increase i by 1
        Determine $K^{(i)} \in \mathcal{P}_c(K)$ such that J is minimized
        by $K^{(i)}$ for (fixed) $U^{(i-1)}$
        Determine $U^{(i)}$ according to theorem 1.12
    UNTIL $\|U^{(i-1)} - U^{(i)}\| \leq \varepsilon$
END FOR

change in the prototypes $K^{(i)}$ is less than ε instead of the change in the membership matrix $U^{(i)}$.

It should be stressed here that algorithms 1 and 2 are only *examples* for algorithms optimizing the objective functions 1.7 and 1.9. Other algorithms for the optimization of these models are hybrid schemes (relaxation) [47], genetic algorithms [1, 54], reformulation [42], and artificial life [86]. The structure of the alternating optimization scheme as in algorithms 1 and 2, however, is well established, and is also used in the cluster estimation algorithms described in chapter 6.

1.7 What to do when the number of clusters is unknown

A disadvantage of the algorithms described in the previous section is that the number c of clusters has to be known in advance. In many applications, this knowledge is not available. In this case, the results of the analysis with different numbers of clusters have to be compared with each other on the basis of another quality function, in order to find an optimal partition with respect to the new objective function. There are two very simple basic techniques. Let D be a data space and let each cluster be uniquely characterized by an element of the set K. We consider a data set $X \subseteq D$ for which we want to determine an optimal cluster partition without knowing the number of clusters in advance.

- The first possibility consists of the definition of a validity function which evaluates a complete cluster partition. An upper bound $c_{\max}$ of the number of clusters has to be estimated, and a cluster analysis corresponding to algorithm 1 or 2 has to be carried out for each $c \in \{2, 3, ..., c_{\max}\}$. For each partition, the validity function now provides a value such that the results of the analysis can be compared indirectly. Possible criteria for the optimal partition may be, for example, a maximal or minimal quality measure. The objective function J according to (1.7) is only partly suited as a validity function since it is monotonously decreasing for an increasing c. The more clusters which are allowed, the smaller the objective function will be, until finally each datum is assigned a cluster of its own and J reaches the value 0. With monotonous criteria like this, the optimal partition can alternatively be determined by the point of maximum curvature of the validity function in dependence of the number of clusters. (Cf. section 7.1 on global validity measures.)

- The second possibility consists of the definition of a validity function that evaluates individual clusters of a cluster partition. Again, an upper bound $c_{\max}$ of the number of clusters has to be estimated and a cluster analysis has to be carried out for $c_{\max}$. The resulting clusters of the analysis are now compared to each other on the basis of the validity function, similar clusters are joined to one cluster, very bad clusters are eliminated. By these operations, the number of clusters is reduced. Afterwards, an analysis according to algorithm 1 or 2 is carried out again with the remaining number of clusters. This procedure is repeated until there is an analysis result containing no similar clusters and no *bad* clusters, with respect to the validity function. (Cf. section 7.2 on local measures of quality.)

Both techniques assume that the results of the analysis are optimal for the respective number of clusters. However, if the cluster analysis provides just locally optimal partitions, only these will be evaluated by the validity function. Thus, an *optimal* cluster partition that is found in this way is possibly only *locally* optimal after all.

In the literature, numerous proposals for validity measures can be found. However, they are almost exclusively related to probabilistic partitions. With a possibilistic clustering, bad partitions, which classify almost all data as outliers, will get a high validity value, because they correctly approximate the few remaining data. If the data contains much high noise so that a possibilistic cluster analysis has to be carried out, one should consider not only the quality of the clusters but also the proportion of the classified data in the set of data. This is especially difficult with quality measures for individual clusters according to the second method. Therefore, in this case it is recommendable to apply the first technique.

The determination of the number of clusters c on the basis of validity measures like the ones mentioned above will be discussed in more detail in chapter 7.

Chapter 2

Classical Fuzzy Clustering Algorithms

Based on the objective functions (1.7), (1.9), and (1.14) several fuzzy models with various distance measures and different prototypes were developed, to which the alternating optimization schemes described in the previous chapter can be applied. The corresponding algorithms are, on the one hand, classical techniques for the recognition of classical cumulus-like clusters (solid clustering, chapter 2), and on the other hand, more recent techniques for the recognition of straight lines (linear clustering, chapter 3), contours of circles, ellipses and parabolas (shell clustering 4) or contours with polygonal boundaries (chapter 5). All techniques are introduced with both their probabilistic versions corresponding to algorithm 1 and their possibilistic versions corresponding to algorithm 2. Of course, the noise clustering method is also applicable to these techniques, taking the changes mentioned on page 30 into account. Clustering algorithms which abandon the objective function and thus the alternating optimization scheme are presented in chapter 6.

As we have already seen in the previous chapter, when we introduced basic generic fuzzy clustering algorithms, an iteration procedure is the basis for our algorithms. Since the exact number of steps to reach convergence is not known a priori, these algorithms are not always well suited for real time applications. Some algorithms that perform particularly extensive computations even seem to be extremely unsuitable for that purpose. However, the decision to apply such a technique strongly depends on the respective experimental setting and it should be considered in any case. In the case of target tracking, for instance, initializing each iteration procedure with the clustering result of the previous one would speed up convergence significantly. In that case, the previous position of a cluster is a good approximation of the following position so that a faster convergence can be expected. Since the number of iteration steps of all iteration procedures strongly depends on their initialization, good results can possibly be achieved in this way, even if another iteration with a random initialization would not have been able to meet the real time constraints.

For problems that do not have to be solved in real time, the application of the methods introduced in the following can definitely be recommended. They are advantageous because of their robustness and their low usage of storage capacities compared to other techniques, such as Hough transformations [44, 45].

Before we dive into the details of the individual techniques, we have to make a brief remark concerning the evaluation of cluster partitions. For two-dimensional data sets, the intuitive, and usually commonly agreed judgement of the data is the inspection by eye and this often sets the standard for the quality of a cluster partition. For higher dimensional data, where even humans cannot recognize an unambiguous cluster partition, one is content with some numerical characteristic data that evaluates the quality of a partition in a strictly mathematical manner. In image or pattern recognition, there is usually no doubt about the *correct* partition; anyone would divide the circles, straight lines or rectangles at hand in the same way. Unfortunately, there is no objective mathematical measure for the partition made that way, such that a program could evaluate its own partitions according to human points of view. Therefore, we speak of *the intuitive partition* when we refer to the partition as it would be made by humans. Without giving a concrete definition of a *good* or *bad* partition, however, these notions remain imprecise and sometimes misleading. If they appear despite these facts, without any further explanation, they refer to that *intuitive partition* that cannot easily be formalized, but the reader can easily comprehend it for each figure.

The techniques introduced in this section deal exclusively with the partition of data into *full* or *solid* clusters (clouds of points, cumulus-like clusters, solid clustering). At the same time, these algorithms represent the beginnings of fuzzy clustering with the methods introduced in the previous chapter. Although we restrict our examples to two-dimensional data sets for illustration purposes, we have to emphasize that the algorithms are also well suited for higher dimensional data.

Especially in image processing, we will often need the basic following algorithms for the initialization of more complicated algorithms.

2.1 The fuzzy c-means algorithm

The development of the fuzzy c-means algorithm (FCM) [31, 8] was the birth of all clustering techniques corresponding to algorithm 1. The first version by Duda and Hart [30] performed a *hard* cluster partition corresponding to definition 1.5 (hard c-means or hard ISODATA algorithm). In order to treat data belonging to several clusters to the same extent in an appropriate manner, Dunn [31] introduced a fuzzy version of this algorithm. It was generalized once more – producing the final version – by Bezdek [8] and his introduction of the *fuzzifier m*. The resulting fuzzy c-means algorithm recognizes spherical clouds of points in a p-dimensional space. The clusters are assumed to be of approximately the same size. Each cluster is represented by its centre. This representation of a cluster is also called a prototype, since it is often regarded as a representative of all data assigned to the cluster. As a measure for the distance, the Euclidean distance between a datum and a prototype is used.

For implementation of the technique corresponding to algorithm 1 or 2, the choice of the optimal cluster centre points for given memberships of the data to the clusters has to be provided. This happens corresponding to theorem 2.1 in the form of a generalized mean value computation, from which the algorithm has also got its name, fuzzy c-*means*. The letter c in the name of the algorithm stands for the number of clusters, for example, with four clusters, it is the fuzzy 4-means. This way of speaking is not strictly kept up, mostly the c remains untouched. It is only supposed to clarify that the algorithm is intended for a fixed number of clusters, i.e. it does not determine that number. The question of determining the number of clusters is discussed in detail in chapter 7.

Theorem 2.1 (Prototypes of FCM) *Let* $p \in \mathbb{N}$, $D := \mathbb{R}^p$, $X = \{x_1, x_2, \ldots, x_n\} \subseteq D$, $C := \mathbb{R}^p$, $c \in \mathbb{N}$, $R := \mathcal{P}_c(C)$, J *corresponding to* (1.7) *with* $m \in \mathbb{R}_{>1}$ *and*

$$d : D \times C \to \mathbb{R}, \ (x,p) \mapsto ||x - p||.$$

If J *is minimized with respect to all probabilistic cluster partitions* $X \to F(K)$ *with* $K = \{k_1, k_2, \ldots, k_c\} \in R$ *and given memberships* $f(x_j)(k_i) = u_{i,j}$ *by* $f : X \to F(K)$, *then*

$$k_i = \frac{\sum_{j=1}^n u_{i,j}^m x_j}{\sum_{j=1}^n u_{i,j}^m} \tag{2.1}$$

holds.

Proof: The probabilistic cluster partition $f : X \to F(K)$ shall minimize the objective function J. Then, all directional derivatives of J with respect to $k_i \in K$, $i \in \mathbb{N}_{\leq c}$ are necessarily 0. Thus, for all $\xi \in \mathbb{R}^p$ with $t \in \mathbb{R}$

$$
\begin{aligned}
0 &= \frac{\partial}{\partial k_i} \sum_{j=1}^{n} \sum_{l=1}^{c} u_{l,j}^{m} ||x_j - k_l||^2 \\
&= \sum_{j=1}^{n} u_{i,j}^{m} \frac{\partial}{\partial k_i} ||x_j - k_i||^2 \\
&= \sum_{j=1}^{n} u_{i,j}^{m} \lim_{t \to 0} \frac{||x_j - (k_i + t\xi)||^2 - ||x_j - k_i||^2}{t} \\
&= \sum_{j=1}^{n} u_{i,j}^{m} \lim_{t \to 0} \frac{1}{t} \Big(((x_j - k_i) - t\xi)^\top ((x_j - k_i) - t\xi) - \\
&\qquad\qquad (x_j - k_i)^\top (x_j - k_i) \Big) \\
&= \sum_{j=1}^{n} u_{i,j}^{m} \lim_{t \to 0} \frac{-2t(x_j - k_i)^\top \xi + t^2 \xi^\top \xi}{t} \\
&= -2 \sum_{j=1}^{n} u_{i,j}^{m} (x_j - k_i)^\top \xi,
\end{aligned}
$$

and it follows that

$$
\frac{\partial}{\partial k_i} J = 0
$$

$$
\Leftrightarrow \sum_{j=1}^{n} u_{i,j}^{m} (x_j - k_i) = 0
$$

$$
\Leftrightarrow k_i = \frac{\sum_{j=1}^{n} u_{i,j}^{m} x_j}{\sum_{j=1}^{n} u_{i,j}^{m}}.
$$

■

In principle, when we choose $m = 1$, the fuzzy c-means algorithm is a generalization of its historical predecessor, the hard c-means algorithm, where the prototypes are computed by the same formula: the memberships are assigned only the "hard" values 0 or 1, though. A datum is assigned to the cluster with the smallest distance. However, $m = 1$ is not allowed for the fuzzy c-means algorithm, since it would lead to a division by zero in (2.1).

In image processing, the FCM algorithm can not be applied to the recognition of shapes but for the determination of positions. In computer-controlled planning of routes in warehouses, vehicles are sometimes equipped with signal lamps on the four corners of the vehicle. A camera on the ceiling of the warehouse captures the whole hall. With the help of an image processing software, the signals of the lamps are then filtered out. Using the FCM algorithm, a determination of the positions of the lamps could be carried out, whose coordinates are provided by the prototypes. With a possibilistic version, this procedure would be quite insensitive to noise data in the data sets.

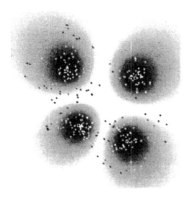
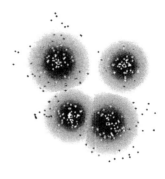

Figure 2.1: FCM analysis   Figure 2.2: P-FCM analysis

Figure 2.1 shows an example data set in $\mathbb{R}^2$ and the result of the fuzzy c-means with randomly initialized prototypes. The cluster partition is indicated by the shading. For each point in the picture plane, the membership to the closest prototype was used for the intensity. Here, high memberships get a dark colour, low memberships a light grey colour. Memberships smaller than $\frac{1}{2}$ are not represented for reasons of better clarity. The data themselves are marked with small black or white crosses in order to clearly distinguish them from the respective background. The clusters' centre points are indicated by small squares.

Note the quite slowly decreasing memberships on those sides where no neighbouring clusters are. In these directions, the memberships are not restricted by the small distances to a different cluster but by the growing distance to all clusters. For possibilistic memberships, this effect does not occur since only the distance to a cluster itself influences the membership to this cluster. The result of a possibilistic clustering is shown in figure 2.2. In comparison to figure 2.1, the lower two prototypes moved closer together, which reflects the intuitive cluster centres a little better. The somewhat

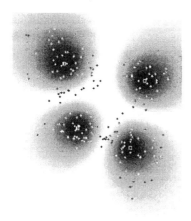
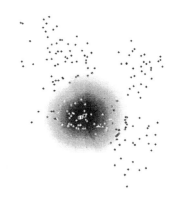

Figure 2.3: FCM analysis   Figure 2.4: P-FCM analysis

greater distance of the prototypes in the probabilistic case, where one is below the other, follows from the tendency of the prototypes to repel each other. All data assigned to a prototype draw it in their direction with a *force* proportional to the membership and the distance measure. If forces from one direction are hardly effective, for instance, because a different cluster is located there and the memberships of these data are therefore low, the prototype automatically follows the forces from the other direction. For a possibilistic clustering, the relationships between different clusters are not important, so that the effect mentioned above does not occur here. Instead, another phenomenon can be observed: if one deals with a set of data for which the cluster centres are not very distinct because clear point accumulations are missing, this has hardly any effect on the probabilistic fuzzy c-means as shown in figure 2.3. The balance of the forces leads to a very similar partition. However, this is completely different for the possibilistic partition 2.4. Closely located point accumulations automatically have a high membership because of their small distance, even if they already belong to a different cluster. The forces are no longer determined by two (sometimes) opposite factors but by the distance measure only. If there is not enough data for each cluster which produces a high force because of its small distance (and thus its high membership), and can in this way fix the cluster, it wanders apparently randomly in each iteration step to the position where most of the data are located within the η_k environment. In these cases, the results of the algorithm can be called useless. In some cases, two, three or even all of the prototypes melt together in one accumulation of points, and the majority of data remains unclassified. For the possibilistic fuzzy c-means, the clusters (that have still to be divided) should have distinct centres of gravity in order to prevent results as in figure 2.4. The noise clustering approach is a good alternative in such cases.

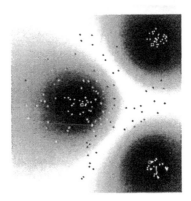

Figure 2.5: FCM analysis

The fuzzy c-means algorithm reaches its limits for clusters of different shapes, sizes and densities. Figure 2.5 shows a case like that: a large cloud of points is approximately located in the middle of the image, flanked by two smaller ones at the edges. In comparison to the intuitive partition, the prototypes have wandered a certain distance away; the prototypes of the small clusters are almost outside the affiliated clouds of points. The outer data of the large cloud of points are close enough to the small clusters in order to become under their influence. They are assigned to two clusters to about the same extent, and they pull the prototypes of the smaller clusters a certain distance towards them. From the perspective of the minimization task, that result was expected; the intuitive idea in image recognition, however, includes the size of the shapes, and implicitly evaluates the distances within a large cloud of points as less strong than those within a small cloud of points.

Figure 2.6: FCM analysis

Another case – different shapes, but the same size of clusters – is shown in figure 2.6. Here, a strong deviation from spherical clusters is sufficient to deceive the FCM algorithm; it is not necessary that each cluster has a different shape. If we had the intuitive partition in the upper and lower ellipses available as an initialization, the fuzzy c-means would perhaps keep that initialization as the final result. The points of the upper ellipse would be closer to the upper prototype than to the lower one and vice versa. The points that are located at a long distance to the left and to the right from the prototypes would neutralize their effects. This way, an equilibrium could be produced, although it will be very unstable. However, as soon as the two prototypes deviate a little from each other in the vertical axis, all points to the left of the ellipses' centres are closer to one of the prototypes and all points to the right are closer to the other prototype. Thus, one prototype is forced to minimize the distance to the left data, while the other prototype minimizes the right-hand side.

With regard to the objective function that has to be minimized, however, the actual result is better than the intuitive partition. While the intuitive partition resulted in maximal distances from the horizontal radii of the ellipses, they are now shrunk to the smaller vertical radii. A general statement such as *"The fuzzy c-means produces in this case a bad partition."* is highly doubtful, because the minimization task at hand was well solved – it was only our model specification for the analysis that was inadequate.

In order to solve the last example by the fuzzy c-means algorithm in the way suggested by intuition, the norm used in the algorithm has to be changed. An arbitrary, positive definite and symmetric matrix $A \in \mathbb{R}^{p \times p}$ induces a scalar product by $< x, y > := x^\top A y$ and a norm by $\|x\|_A := \sqrt{< x, x >}$, respectively. If the standard norm is replaced by an A-norm in the fuzzy c-means algorithm, nothing changes regarding the validity of theorem 2.1 because special properties of the norm were not used in its proof. The matrix $A = \begin{pmatrix} \frac{1}{4} & 0 \\ 0 & 4 \end{pmatrix}$ provides

$$\left\| \begin{pmatrix} x \\ y \end{pmatrix} \right\|_A^2 = \begin{pmatrix} x & y \end{pmatrix} \begin{pmatrix} \frac{1}{4} & 0 \\ 0 & 4 \end{pmatrix} \begin{pmatrix} x \\ y \end{pmatrix} = \begin{pmatrix} \frac{1}{4}x^2 \\ 4y^2 \end{pmatrix} = \begin{pmatrix} \left(\frac{x}{2}\right)^2 \\ (2y)^2 \end{pmatrix}.$$

This corresponds to a deformation of the unit circle to an ellipse with a double diameter in the x-direction and half a diameter in the y-direction. Thus, if there is a priori knowledge about the data, even elliptic clusters can be recognized by the fuzzy c-means algorithm, provided all clusters have the same shape.

One of FCM's advantages is its simplicity leading to low computation time. In practice, a few iteration steps already provide good approximations to the final solution so that five FCM steps may serve as an initialization for further techniques.

2.2 The Gustafson-Kessel algorithm

By replacing the Euclidean distance by another metric induced by a positive definite, symmetric matrix in the fuzzy c-means algorithm, ellipsoidal clusters could also be recognized, instead of only spherical ones. However, the fuzzy c-means algorithm is not suited for an automatic adaptation for each individual cluster. An algorithm designed for this task was proposed by Gustafson and Kessel (GK) [41].

In comparison with the fuzzy c-means algorithm, in addition to the cluster centres each cluster is characterized by a symmetric and positive definite matrix A. This matrix induces for each cluster a norm of its own $||x||_A := \sqrt{x^\top A x}$. Here, it has to be taken into account that the distances could become arbitrarily small with an arbitrary choice of the matrices. In order to avoid a minimization of the objective function by matrices with (almost) zero entries, we require a constant volume of clusters by $\det(A) = 1$. Thus, only the cluster shapes are variable now, but not the clusters' sizes. Gustafson and Kessel permit different sizes of clusters, too, by introducing a constant value ϱ for each matrix A, and they generally demand $\det(A) = \varrho$. However, the choice of the constants requires a priori knowledge about the clusters again.

The determination of the prototypes is carried out, corresponding to

Theorem 2.2 (Prototypes of GK) *Let* $p \in \mathbb{N}$, $D := \mathbb{R}^p$, $X = \{x_1, x_2, \ldots, x_n\} \subseteq D$, $C := \mathbb{R}^p \times \{A \in \mathbb{R}^{p \times p} \mid \det(A) = 1$, A *symmetric and positive definite*$\}$, $c \in \mathbb{N}$, $R := \mathcal{P}_c(C)$, J *corresponding to (1.7) with* $m \in \mathbb{R}_{>1}$ *and*

$$d^2 : D \times C \to \mathbb{R}, \ (x, (v, A)) \mapsto (x - v)^\top A(x - v).$$

If J *is minimized by* $f : X \to F(K)$ *with respect to all probabilistic cluster partitions* $X \to F(K)$ *with* $K = \{k_1, k_2, \ldots, k_c\} \in R$ *and given memberships* $f(x_j)(k_i) = u_{i,j}$, *with* $k_i = (v_i, A_i)$, *then*

$$v_i = \frac{\sum_{j=1}^n u_{i,j}^m x_j}{\sum_{j=1}^n u_{i,j}^m} \tag{2.2}$$

$$A_i = \sqrt[p]{\det(S_i)} S_i^{-1} \tag{2.3}$$

$$S_i \;=\; \sum_{j=1}^{n} u_{i,j}^{m}(x_j - v_i)(x_j - v_i)^{\top} \tag{2.4}$$

holds.

Proof: In order to take the constraints of constant cluster volumes, c Lagrange multipliers λ_i, $i \in \mathbb{N}_{\leq c}$, have to be introduced, such that the objective function results in $J = \sum_{j=1}^{n} \sum_{i=1}^{c} u_{i,j}^{m} \|x_j - v_i\|_{A_i}^2 - \sum_{i=1}^{c} \lambda_i(\det(A_i) - 1)$. The probabilistic cluster partition $f : X \to F(K)$ shall minimize the objective function J. Once again, all directional derivatives of J with respect to k_i, $i \in \mathbb{N}_{\leq c}$ are necessarily 0. Let $k_i = (v_i, A_i) \in K$. For the cluster centre positions v_i we obtain again equation (2.2) as in theorem 2.1, because the considerations there were independent of the norm.

Taking the derivative of J with respect to the matrix A_i, we go beyond the restriction to symmetric, positive definite matrices with determinant 1, and instead we consider all regular matrices of $\mathbb{R}^{p \times p}$. After that, we can calculate the partial derivatives with respect to all matrix elements. (The set of the invertible matrices is itself open as the inverse image of the open set $\mathbb{R}_{>0}$ under the continuous mapping assigning to each matrix its determinant, and therefore it is differentiable in all directions.) We have $\nabla x_j^{\top} A_i x_j = x_j x_j^{\top}$ and $\nabla \det(A_i) = \det(A_i) A_i^{-1}$. Since the gradient has to be zero for minimzing the objective function, we obtain

$$0 = \nabla J = \left(\sum_{j=1}^{n} u_{i,j}^{m}(x_j - v_i)(x_j - v_i)^{\top} \right) - \lambda_i \det(A_i) A_i^{-1}.$$

The derivatives with respect to the Lagrange multipliers lead to the constraints $\det(A_i) = 1$ for $i \in \mathbb{N}_{\leq c}$. With the notation S_i from the theorem, altogether we obtain

$$S_i = \lambda_i A_i^{-1}. \tag{2.5}$$

We denote the identity matrix by $I \in \mathbb{R}^{p \times p}$ and, from that together with the invertibility of A_i, it follows that

$$S_i A_i = \lambda_i I$$

holds. Taking the determinant of this equation leads to

$$det(S_i A_i) = \lambda_i^{p}, \quad \text{and thus} \quad \lambda_i = \sqrt[p]{\det(S_i)\det(A_i)} = \sqrt[p]{\det(S_i)}.$$

If we replace the Lagrange multiplier in equation (2.5) by this expression, we obtain equation (2.3):

$$A_i = \sqrt[p]{\det(S_i)} S_i^{-1}.$$

It remains to be shown that A_i also satisfies the constraint of a positive definite and symmetric matrix, that we had neglected in the beginning. In order to do that, we have to assume that there are p linearly independent vectors $\xi \in \mathbb{R}^p$ in the data set. Then, the matrices $\xi \xi^\top$ are symmetric and positive semi-definite and also their weighted sum and thus A_i is symmetric and positive definite. (The invertibility that we had assumed is therefore also satisfied.) Because of $\det(A^{-1}) = \frac{1}{\det(A)}$, we also have

$\det(A_i) = \det(S_i^{-1}) \left(\sqrt[p]{\det(S_i)} \right)^p = \frac{1}{\det(S_i)} \det(S_i) = 1$. Since f represents a minimum of the objective function on the set of all regular matrices, and furthermore A_i is positive definite with determinant 1, f especially represents a minimum on the restricted space of matrices. ∎

Figure 2.7: GK analysis

Instead of the matrices S_i, Gustafson and Kessel use so-called fuzzy covariance matrices $\dfrac{\sum_{j=1}^n u_{i,j}^m (x_j - v_i)(x_j - v_i)^\top}{\sum_{j=1}^n u_{i,j}^m}$. However, the factor $\dfrac{1}{\sum_{j=1}^n u_{i,j}^m}$ is not relevant for the result, because the matrices are scaled to the unit determinant. The importance of these fuzzy covariance matrices will be clarified in the context of remark 2.3 from section 2.3.

The target applications of the Gustafson-Kessel algorithm are similar to those of the fuzzy c-means algorithm. Because of the adaptation of the distance function to the clusters, the results for non-spherical clusters correspond better to the intuitive partitions.

The data sets from figures 2.1, 2.2, 2.3, and 2.4 are clustered by the Gustafson-Kessel algorithm with results similar to those of the fuzzy c-means algorithm. For the Gustafson-Kessel algorithm, the clusters have the shapes of ellipses that fit the data a little bit better than the spherical shapes resulting from the fuzzy c-means algorithm. The problem of point accumulations in the cluster centre still remains in the case of possibilistic clustering with GK.

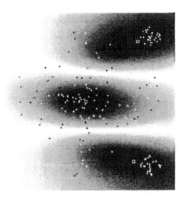

Figure 2.8: GK analysis

Let us now consider the example data sets that caused problems for the fuzzy c-means algorithm. The two long-stretched ellipses from figure 2.6 are clustered by the Gustafson-Kessel algorithm as expected, in accordance with the intuition, as shown in figure 2.7. However, this data set is not a real challenge for the algorithm; both clusters have the same elliptical shape of an ellipse after all, such that the fuzzy c-means algorithm would have been sufficient, provided that we would have used an adequate modified metric induced by a symmetric and positive definite matrix. Before we consider example data with different cluster shapes, we take a look at the second data set which caused problems for the fuzzy c-means algorithm. Figure 2.8 shows two small clusters flanking a large cluster in the middle (cf. figure 2.5). Here, the Gustafson-Kessel algorithm recognized three elliptic clusters, which did not result in a better approximation of the intuitive partition, though. On the contrary, the prototypes of the small clusters were influenced even more by the edge data of the large cluster, so that these cluster centres are moved even further away from their intuitively correct position. It is only the prototype of the large cluster that came closer to the intuitive position.

Figure 2.9 shows the memberships of a probabilistic Gustafson-Kessel cluster partition of four ellipses arranged in a square. The data are deliberately noisy in the area of the lower right corner. These noise data pull the prototype out of the ellipse located nearby, and deform the cluster shape to a certain degree. The other ellipses are well recognized. Figure 2.10 shows the same data set after a possibilistic analysis. Noise data were widely recognized as such, and the cluster on the bottom right was focused exactly to the affiliated ellipse. On the other hand, the cluster on the bottom left also classified some of the left edge points as noise data after this cluster had

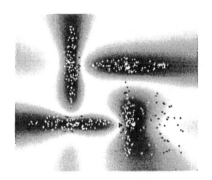

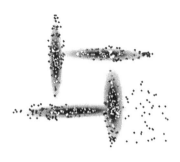

Figure 2.9: GK analysis   Figure 2.10: P-GK analysis

wandered a little to the right-hand side. Similar changes can be noticed for the upper clusters, although they are not that strong. On the whole, the partition corresponds quite well to the intuition.

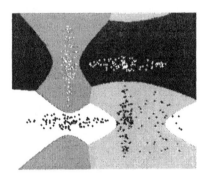

Figure 2.11: Regions covered by the clusters

Figure 2.11 shows the same data set once more. This time the grey values indicate to which cluster a datum would be assigned, based on the maximum membership of the datum. In this way it becomes obvious that some of the noise points on the extreme right-hand side were unexpectedly assigned to the bottom left cluster. Since these points in figure 2.9 are already in the area with increasing membership values on the very right-hand side, we can conclude that the membership degree is greater than $\frac{1}{2}$. Because of the rather high distance to their prototype, they are the reason for the lower left cluster to be positioned a little more to the right than intuitively expected, as in figure 2.9. The high membership of the extreme right noise data to the left cluster, which must seem rather odd, can be explained by the norm that is introduced in the Gustafson-Kessel algorithm. The matrices A_i stretch the distances along the ellipses' axes to a different extent. For the ellipse on the bottom left, the horizontal distances are shrunk so that far distant points with respect to the Euclidean norm now come very close. The opposite is the case for the ellipse on the bottom right. In the horizontal axis, the distances become greater than with the Euclidean norm. For this reason, all data to the *right* of the bottom right ellipse, beginning from a certain distance, will be assigned to the bottom *left* ellipse. Noise data that are located there can have a surprising effect, because they affect an ellipse that was seemingly not involved. We will observe similar unwanted consequences of many other algorithms that result from deviations from the Euclidean distance. However, the effect is much less obvious when the noise data are distributed over the whole image and do not occur so intensively in one spot.

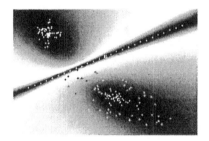

Figure 2.12: GK analysis   Figure 2.13: P-GK analysis

Two further example data sets for the Gustafson-Kessel algorithm are shown in figures 2.12 and 2.13. They represent a straight line, a spherical and an elliptic cluster. For the probabilistic (figure 2.12) as well as the possibilistic cluster partition (figure 2.13), the algorithm provides results that exactly meet the wanted and intuitive partition. The good detection of the straight line by an extremely long-stretched ellipse is remarkable (cf. also chapter 3).

If the data can be possibilistically clustered as in the previous example, the extension factors η of the clusters can be used to actually recognize the shapes from the image. Position and orientation can be obtained from the cluster centre and the matrix. The positive definite, symmetric matrix defines a rotation and a stretching. A rough estimation of the size of the cluster can then be obtained from the extension factor η. A photo of a stick, a large elliptic and a spherical board – represented by a data set as in figure 2.13 – then could be recognized by the Gustafson-Kessel algorithm. However, the shapes would be equally as densely covered with data after a conventional image post-processing of the picture. As already mentioned, an accumulation of data in the area of the centre is desirable for a stable execution of the possibilistic algorithms. So the quality of the result obtained with this technique strongly depends on the available data.

In comparison to the fuzzy c-means algorithm the computational costs of the Gustafson-Kessel algorithm are much higher. In each iteration step $n \cdot c$ matrices $\mathbb{R}^{p \times p}$ are computed, added up to c matrices and inverted afterwards. Furthermore, c determinants are computed. This results in a distinctly longer computing time, especially for higher dimensional data sets. In any case, it may be useful to initialize the cluster centres of the GK prototypes with the resulting prototypes of a fuzzy c-means run in order to reach faster convergence and reduce the number of iteration steps.

2.3 The Gath-Geva algorithm

The algorithm by Gath and Geva (GG) [37] is an extension of the Gustafson-Kessel algorithm that also takes the size and density of the clusters into account. Gath and Geva interpret the data as realizations of p-dimensional, normally distributed random variables. If we consider for a moment an unambiguous assignment of the n data x_j, $j \in \mathbb{N}_{\leq n}$, to the c normal distributions N_i, $i \in \mathbb{N}_{\leq c}$, this corresponds to the hard memberships $u_{i,j} \in \{0, 1\}$. In that case, statistics provides the estimator

$$m_i = \frac{\sum_{j=1}^{n} u_{i,j} x_j}{\sum_{j=1}^{n} u_{i,j}}$$

for the expected value of the ith normal distribution and

$$C_i = \frac{\sum_{j=1}^{n} u_{i,j} (x_j - m_i)(x_j - m_i)^\top}{\sum_{j=1}^{n} u_{i,j}}$$

for the covariance matrix. The similarity of these results with the computation instructions for the Gustafson-Kessel algorithm suggests a generalization of the results from probability theory to the case of fuzzy memberships of the data to the normal distributions.

Gath and Geva assume that the normal distribution N_i with expected value v_i and covariance matrix A_i is chosen for generating a datum with a priori probability P_i. The unnormalized a posteriori probability (likelihood) for a datum x_j to be generated by the normal distribution N_i is thus

$$\frac{P_i}{(2\pi)^{p/2}\sqrt{\det(A_i)}} \exp(-\frac{1}{2}(x_j - v_i)^\top A_i^{-1}(x_j - v_i)).$$

The distance function for the Gath-Geva algorithm is chosen to be indirectly proportional to this (unnormalized) a posteriori probability. Therefore, we use the reciprocal value of this formula and neglect the constant factor $(2\pi)^{p/2}$.

In contrast to the fuzzy c-means and the Gustafson-Kessel algorithm, the Gath-Geva algorithm is not based on an objective function, but is a fuzzification of statistical estimators. We obtain the following iteration instructions for the Gath-Geva algorithm.

Remark 2.3 (Prototypes of GG)
Let $p \in \mathbb{N}$, $D := \mathbb{R}^p$, $X = \{x_1, x_2, \ldots, x_n\} \subseteq D$, $C := \mathbb{R}^p \times \{A \in \mathbb{R}^{p \times p} \mid A \text{ positive definite and symmetric}\} \times \mathbb{R}$, $c \in \mathbb{N}$, $R := \mathcal{P}_c(C)$, J corresponding to (1.7) with $m \in \mathbb{R}_{>1}$ and

$$d^2 : D \times C \rightarrow \mathbb{R},$$
$$(x, (v, A, P)) \mapsto \frac{1}{P} \sqrt{\det(A)} \exp\left(\frac{1}{2}(x - v)^\top A^{-1}(x - v)\right).$$

In order to approximately minimize J with respect to all probabilistic cluster partitions $X \rightarrow F(K)$ with $K = \{k_1, k_2, \ldots, k_c\} \in R$ and given memberships $f(x_j)(k_i) = u_{i,j}$ by $f : X \rightarrow F(K)$, the parameters $k_i = (v_i, A_i, P_i)$ of the normal distribution N_i should be chosen as follows:

$$v_i = \frac{\sum_{j=1}^n u_{i,j}^m x_j}{\sum_{j=1}^n u_{i,j}^m}$$

$$A_i = \frac{\sum_{j=1}^n u_{i,j}^m (x_j - v_i)(x_j - v_i)^\top}{\sum_{j=1}^n u_{i,j}^m}$$

$$P_i = \frac{\sum_{j=1}^n u_{i,j}^m}{\sum_{j=1}^n \sum_{l=1}^c u_{l,j}^m}.$$

The generalization of the statistical estimators for the expected value and the covariance matrix directly leads to the computation instructions for the cluster centres v_i and the fuzzy covariance matrices A_i. P_i estimates the a priori probability for the membership of an arbitrary datum to the normal distribution N_i corresponding to the principle "*number of data in cluster i divided by the total number of data*". The normalized a posteriori probability specifies the probability that a *given* datum was generated by the ith normal distribution. The distance function $d^2(x_j, k_i)$ is chosen indirectly proportional to the (unnormalized) a posteriori probability (likelihood), since a small distance means a high probability and a large distance means a low probability for the membership.

If we were to apply for the Gath-Geva algorithm the same technique as for the fuzzy c-means and the Gustafson-Kessel algorithms, i.e. consider the derivatives of the function J (1.7) in order to obtain equations for the prototypes, the resulting system of equations could not be solved analytically. In this sense, the Gath-Geva algorithm is a good heuristics on the basis of an analogy with probability theory.

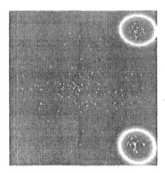

Figure 2.14: GG analysis

The Gath-Geva algorithm clusters all examples mentioned in the sections on the fuzzy c-means and Gustafson-Kessel algorithms very well in the sense of the respective intuitive partition. Even the data from figure 2.8 are clustered as desired – a result that could not be obtained by applying the fuzzy c-means or the Gustafson-Kessel algorithms. The Gath-Geva algorithm recognizes the middle cluster with its low data density and large extension, and the cluster is well separated from the two satellite clusters as shown in figure 2.14. The memberships present a somewhat strange picture. Because of the occurrence of the exponential function within the distance function, all distances are more or less divided into the two ranges, close and remote. From a certain distance on, the distances increase so much because of the exponential function, that there are hardly any values other than 0 and 1 for the memberships near the clusters. In the areas of transitions from one cluster to another, the memberships change very rapidly from 0 to 1, and vice versa. This change is indicated in the figure by the bright circles. Thus, the memberships induce a very strict separation of the three clusters, approximately in the same way humans would assign the data to the clusters. (The grey levels chosen are a little lighter in figures 2.14 and 2.15 in comparison to the previous examples. The large grey areas have the maximum intensity in the original representation.)

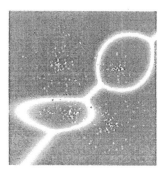

Figure 2.15: GG analysis

Similarly good results are shown in figure 2.15. This data set was already analysed with the fuzzy c-means algorithm in figures 2.1 and 2.2. Here also, the partition of the data is almost *hard*. While there were clearly more data whose maximal membership was below $\frac{1}{2}$ in the partition obtained from the fuzzy c-means algorithm, their number was considerably decreased with the algorithm by Gath and Geva. The clusters' shapes did not change much, but the partition is distinctly more concrete because of the memberships. Even with very strongly varying cluster shapes and sizes as in figure 2.16, the algorithm yields very good results. The picture only shows the different regions that are covered by the clusters when data are assigned to a cluster based on their maximum membership degree. The cluster centres, too, are exactly in the centre of the respective cluster; they did not wander because of the assignment of a few edge points, as before.

However, it has to be noted that the clustering algorithms become more sensible to local minima with increasing complexity. It is more likely for the Gath-Geva algorithm to converge in a local minimum than it is for the Gustafson-Kessel algorithm; and the fuzzy c-means algorithm is usually little affected by local minima. For different initializations of the prototypes, the partitions of the Gath-Geva algorithm can be very different. A GK initialization was necessary in addition to an FCM initialization for some of the figures (e.g. 2.16), however, an additional GK initialization caused a worse analysis result for other figures (e.g. 2.15). For the *correct* partition by the Gath-Geva algorithm, the prototypes have to be initialized near the final prototypes. The initialization chosen should be dependent on whether the fuzzy c-means or the Gustafson-Kessel algorithm better meets the desired positions of the prototypes. Unfortunately, this decision cannot always be made without any a priori knowledge.

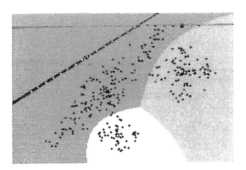

Figure 2.16: Regions covered by the clusters

Furthermore, a possibilistic clustering should not be performed with the Gath-Geva algorithm. Since the algorithm independently adjusts the size of the clusters, the cut-off of the noise data leads to smaller clusters with each iteration step so that the cluster size converges towards 0. This is no surprise because we assumed a probabilistic interpretation of the memberships in remark 2.3. However, the Gath-Geva prototypes can be transformed into Gustafson-Kessel prototypes after the probabilistic GG run. Here, the information about the size of the clusters has to be made available to the Gustafson-Kessel algorithm by a constant factor applied to the cluster norm induced by the corresponding matrix A_i. Now the possibilistic Gustafson-Kessel algorithm could again determine the clusters shape, but not the cluster sizes. In this way, the a priori probability and the extension factors η do not influence each other any more. The degeneration of the clusters is thus avoided. A better alternative is the application of noise clustering in the framework of the Gath-Geva algorithm, since noise clustering maintains a probabilistic interpretation.

For the implementation of the Gath-Geva algorithm, one should be aware of the fact that there can easily occur floating-point overflows because of the exponential function. Even with relatively small Euclidean distances, the Gath-Geva distances become very large. When calculating the memberships, large distances lead to very small values. Therefore, it is perfectly acceptable to use a modified exponential function that provides constant or only linearly increasing values when the arguments face an overflow. The resulting imprecision for higher memberships is scarcely of any importance because of the very small values.

2.4   Simplified versions of the Gustafson-Kessel and Gath-Geva algorithms

The Gustafson-Kessel and Gath-Geva algorithms extend the fuzzy c-means algorithm by the computation of a covariance matrix for each cluster. By means of these covariance matrices, clusters with the shape of ellipses or ellipsoids can be recognized, while the fuzzy c-means algorithm is tailored for circular and spherical clusters, respectively. The covariance matrix encodes a transformed norm that transfers the unit circle or the unit sphere to an ellipse or an ellipsoid, respectively. This means that each covariance matrix encodes a scaling of the axes as well as a rotation. If the covariance matrix is a diagonal matrix, only the axes are scaled and no rotation is performed, so that the unit sphere is transformed to an axes-parallel hyper-ellipsoid.

In the following, we present modifications of the Gustafson-Kessel and Gath-Geva algorithms that permit diagonal matrices only, instead of arbitrary positive definite and symmetric matrices [56, 58]. Even though these versions are less flexible than the original algorithms, they are still more flexible than the fuzzy c-means algorithm, they avoid the inversion of matrices and the computation of determinants, and they are better suited for the generation of fuzzy rules, as we will see in chapter 8.

First we look at the axes-parallel version of the Gustafson-Kessel algorithm (AGK). Analogous to theorem 2.2, the determination of the prototypes of the axes-parallel version of the Gustafson-Kessel algorithm results from

Theorem 2.4 (Prototypes of AGK) *Let* $p \in \mathbb{N}$, $D := \mathbb{R}^p$, $X = \{x_1, x_2, \ldots, x_n\} \subseteq D$, $C := \mathbb{R}^p \times \{A \in \mathbb{R}^{p \times p} \mid \det(A) = 1$, *$A$ diagonal matrix\}, $c \in \mathbb{N}$, $R := \mathcal{P}_c(C)$, J corresponding to (1.7) with $m \in \mathbb{R}_{>1}$ and*

$$d^2 : D \times C \to \mathbb{R}, \ (x, (v, A)) \mapsto (x - v)^\top A(x - v).$$

If J is minimized with respect to all probabilistic cluster partitions $X \to F(K)$ with $K = \{k_1, k_2, \ldots, k_c\} \in R$ and given memberships $f(x_j)(k_i) = u_{i,j}$ with $k_i = (v_i, A_i)$ by $f : X \to F(K)$, then

$$v_i = \frac{\sum_{j=1}^n u_{i,j}^m x_j}{\sum_{j=1}^n u_{i,j}^m} \tag{2.6}$$

$$a_\gamma^{(i)} = \frac{\left(\prod_{\alpha=1}^p \sum_{j=1}^n (u_{i,j})^m (x_{j,\alpha} - v_{i,\alpha})^2\right)^{1/p}}{\sum_{j=1}^n (u_{i,j})^m (x_{j,\gamma} - v_{i,\gamma})^2} \tag{2.7}$$

holds where $a_\gamma^{(i)}$ is the γth diagonal element of the diagonal matrix A_i.

Proof: For the cluster centre vector v_i, we obtain again (2.6) from theorem 2.1, because the considerations were independent of the norm there.

The requirement, that the determinant of the diagonal matrix A_i has to be 1, is equivalent to

$$1 = \prod_{\alpha=1}^p a_\alpha^{(i)}. \tag{2.8}$$

In order to minimize the objective function, we introduce Lagrange multipliers for this constraint so that we obtain the new objective function

$$\sum_{j=1}^n \sum_{i=1}^c u_{i,j}^m \sum_{\alpha=1}^p a_\alpha^{(i)} (x_{j,\alpha} - v_{i,\alpha})^2 - \sum_{i=1}^c \lambda_i \left(\left(\prod_{\alpha=1}^p a_\alpha^{(i)} \right) - 1 \right)$$

Here, $x_{j,\alpha}$ and $v_{i,\alpha}$ denote the αth component of the vectors x_j and v_i, respectively.

The derivative of this objective function with respect to the vector $a^{(i)} = (a_1^{(i)}, \ldots, a_p^{(i)})^\top$ must be zero for a minimum. Let $\xi \in \mathbb{R}^p$ be an arbitrary unit vector. Thus, we obtain

$$\frac{\partial J}{\partial \xi}(a^{(i)}) = \lim_{t \to 0} \frac{J(a^{(i)} + t \cdot \xi_\alpha) - J(a^{(i)})}{t}$$

$$= \lim_{t \to 0} \frac{1}{t} \left(\sum_{\alpha=1}^p t\xi \sum_{j=1}^n (u_{i,j})^m (x_{j,\alpha} - v_{i,\alpha})^2 \right.$$

$$\left. - \lambda_i \sum_{\alpha=1}^p t\xi_\alpha \prod_{\beta=1, \beta\neq\alpha}^p a_\beta^{(i)} + o(t^2) \right)$$

$$= \sum_{\alpha=1}^p \xi_\alpha \sum_{j=1}^n (u_{i,j})^m (x_{j,\alpha} - v_{i,\alpha})^2 - \lambda_i \sum_{\alpha=1}^p \xi_\alpha \prod_{\beta=1, \beta\neq\alpha}^p a_\beta^{(i)}$$

$$= \sum_{\alpha=1}^p \xi_\alpha \left(\sum_{j=1}^n (u_{i,j})^m (x_{j,\alpha} - v_{i,\alpha})^2 - \lambda_i \prod_{\beta=1, \beta\neq\alpha}^p a_\beta^{(i)} \right)$$

$$= 0$$

independent of ξ. Thus, we have

$$\lambda_i \prod_{\beta=1,\,\beta\neq\gamma}^{p} a_\beta^{(i)} = \sum_{j=1}^{n} (u_{i,j})^m (x_{j,\gamma} - v_{i,\gamma})^2 \qquad (2.9)$$

for all $\gamma \in \{1, \ldots, p\}$. Taking the constraint (2.8) into account, the left-hand side of (2.9) may be replaced by $\lambda_i / a_\gamma^{(i)}$ leading to

$$a_\gamma^{(i)} = \frac{\lambda_i}{\sum_{j=1}^{n} (u_{i,j})^m (x_{j,\gamma} - v_{i,\gamma})^2}.$$

Making use of this result in (2.8), we obtain

$$\lambda_i = \left(\prod_{\alpha=1}^{p} \sum_{j=1}^{n} (u_{i,j})^m (x_{j,\alpha} - v_{i,\alpha})^2 \right)^{1/p}$$

and – together with the previous equation – we finally have (2.7). ∎

For modification of the Gath-Geva algorithm, we assume again that the data are realizations of p-dimensional normal distributions. However, we furthermore assume that each of these normal distributions is induced by p independent, one-dimensional normal distributions, i.e. the covariance matrix is a diagonal matrix with the variances of the one-dimensional normal distributions as diagonal elements. We again denote the a priori probability that a datum is generated by the ith normal distribution by P_i. Let $a_\alpha^{(i)}$ be the αth element of the diagonal of the ith covariance matrix A_i. Thus,

$$g_i(x) = \frac{1}{(2\pi)^{p/2}} \cdot \frac{1}{\sqrt{\prod_{\alpha=1}^{p} a_\alpha^{(i)}}} \cdot \exp\left(-\frac{1}{2} \sum_{\alpha=1}^{p} \frac{(x_\alpha - v_{i,\alpha})^2}{a_\alpha^{(i)}} \right)$$

is the probability density function of the ith normal distribution.

As was already the case for the original Gath-Geva algorithm, the probabilistic parameters are estimated with statistical methods. For the a priori probabilities P_i and the expected values v_i, we obtain the same formulae for the modified version. We introduce a fuzzification of the a posteriori probabilities in order to determine the parameters $a_\alpha^{(i)}$. The unnormalized a posteriori probability that *all* data were generated by the ith cluster is

$$\prod_{j=1}^{n} P_i g_i(x_j). \qquad (2.10)$$

However, only those data have to be considered that actually belong to the ith cluster. Hence, we modify (2.10) and obtain

$$\prod_{j=1}^{n} (P_i g_i(x_j))^{(u_{i,j})^m} . \tag{2.11}$$

This formula is known in statistical mixture models [19]. Obviously, this formula becomes

$$\prod_{j:\, x_j \text{ belongs to cluster } i} P_i g_i(x_j)$$

in the case of a hard cluster partition, i.e. $u_{i,j} \in \{0,1\}$. Now we determine the maximum likelihood estimator for the formula (2.11) by choosing the parameter $a_\alpha^{(i)}$ such that the (unnormalized) a posteriori probability reaches a maximum. Instead of directly maximizing (2.11), we apply the logarithm to this formula[1] and obtain

$$F(a_1^{(i)}, \ldots, a_p^{(i)}) = \sum_{j=1}^{n} (u_{i,j})^m \left(\ln(P_i) - \frac{p}{2} \ln(2\pi) - \frac{1}{2} \sum_{\alpha=1}^{p} \ln(a_\alpha^{(i)}) \right.$$

$$\left. - \frac{1}{2} \sum_{\alpha=1}^{p} \frac{(x_{j;\alpha} - v_{i,\alpha})^2}{a_\alpha^{(i)}} \right). \tag{2.12}$$

The partial derivatives of this function with respect to $a_\gamma^{(i)}$ lead to

$$\frac{\partial F(a_1^{(i)}, \ldots, a_p^{(i)})}{\partial a_\gamma^{(i)}} = -\frac{1}{a_\gamma^{(i)}} \cdot \frac{1}{2} \sum_{j=1}^{n} (u_{i,j})^m$$

$$+ \frac{1}{2} \sum_{j=1}^{n} (u_{i,j})^m \frac{(x_{j,\gamma} - v_{i,\gamma})^2}{(a_\gamma^{(i)})^2}. \tag{2.13}$$

In order to minimize (2.12), (2.13) has to be zero so that we obtain

$$a_\gamma^{(i)} = \frac{\sum_{j=1}^{n} (u_{i,j})^m (x_{j,\alpha} - v_{i,\alpha})^2}{\sum_{j=1}^{n} (u_{i,j})^m}$$

as an estimation for the parameter $a_\gamma^{(i)}$.

---

[1] Since the logarithm is a monotonously increasing function, it does not matter whether the maximum of a (positive) function is determined or the maximum of its logarithm.

The (axes-parallel) version of the Gath-Geva algorithm that is restricted to diagonal matrices is thus

Remark 2.5 (Prototypes of AGG) *Let* $p \in \mathbb{N}$, $D := \mathbb{R}^p$, $X = \{x_1, x_2, \ldots, x_n\} \subseteq D$, $C := \mathbb{R}^p \times \{A \in \times \mathbb{R}^{p \times p} \mid A$ *diagonal matrix*, $\det(A) \neq 0\} \times \mathbb{R}$, $c \in \mathbb{N}$, $R := \mathcal{P}_c(C)$, J *corresponding to* (1.7) *with* $m \in \mathbb{R}_{>1}$ *and*

$$d^2 : D \times C \to \mathbb{R},$$

$$(x, (v, A, P)) \mapsto \frac{1}{P} \sqrt{\det(A)} \exp\left(\frac{1}{2}(x - v)^\top A^{-1}(x - v)\right).$$

In order to approximately minimize J *with respect to all probabilistic cluster partitions* $X \to F(K)$ *with* $K = \{k_1, k_2, \ldots, k_c\} \in R$ *and given memberships* $f(x_j)(k_i) = u_{i,j}$ *by* $f : X \to F(K)$, *the parameters* $k_i = (v_i, A_i, P_i)$ *of the normal distribution* N_i *should be chosen as follows:*

$$v_i = \frac{\sum_{j=1}^n u_{i,j}^m x_j}{\sum_{j=1}^n u_{i,j}^m}$$

$$a_\gamma^{(i)} = \frac{\sum_{j=1}^n (u_{i,j})^m (x_{j,\alpha} - v_{i,\alpha})^2}{\sum_{j=1}^n (u_{i,j})^m}$$

$$P_i = \frac{\sum_{j=1}^n u_{i,j}^m}{\sum_{j=1}^n \sum_{l=1}^c u_{l,j}^m}$$

where $a_\gamma^{(i)}$ *is the* γ*th diagonal element of the matrix* A_i.

The versions of the Gustafson-Kessel and the Gath-Geva algorithms that are restricted to diagonal matrices require neither the computation of a determinant nor the inversion of a matrix, so that they need much less computation time, but lose a certain amount of flexibility. Application areas of these algorithms are presented in the chapter on rule generation with fuzzy clustering.

2.5   Computational effort

All algorithms mentioned above were implemented in the programming language C. In order to be able to compare the algorithms, a time index τ was given. This index was formed from the quotient of the computation time t that was needed (in seconds, but used without dimensions) and the product of the number of data n, the number of clusters c, and the steps of the iteration i: $\tau = \frac{t}{c \cdot n \cdot i}$. The time indices for different analyses of the same algorithm are not constant. For ten seconds of computing time with 10,000 data items and 10 clusters we obtained a lower time index than with 1000 data and 100 clusters (same number of iteration steps). The more complicated computations (e.g. matrix inversion) are necessary once per cluster in each iteration step so that the total effort for 100 clusters is of course greater than for 10 clusters. The number of clusters does not vary very much in the examples so that the different analysis results in comparable values. Each given time index is a mean of all indices that were represented in the respective section. The program that was used was not optimized with respect to the execution speed so that the index might be decreased by a more efficient implementation. However, all algorithms are optimized (or not optimized) to almost the same extent, so that the indices should reflect the computational needs of the various algorithms. The expense for the initialization is not included in the time index.

Figures 2.1 – 2.16 contained data sets with 170 up to 455 points. In most cases, it took the FCM algorithm only a few iterations (< 30), and so it remained below one second of computation time. The possibilistic FCM algorithm sometimes needs significantly more iterations; three passes are necessary after all (a probabilistic one and two possibilistic ones), and the η values have to be calculated twice, once after the probabilistic pass and once after the first possibilistic pass.

The GK algorithm needed about 30 steps for the examples, the GG algorithm between 15 and 75 depending on the initialization. The time index was $\tau_{FCM} = 7.02 \cdot 10^{-5}$ for FCM, $\tau_{GK} = 1.50 \cdot 10^{-4}$ for GK, and $\tau_{GG} = 1.78 \cdot 10^{-4}$ for GG. The expensive operations such as exponential functions and matrix inversions, show their effects here in comparison to the simple FCM algorithm.

The precision for termination of $\frac{1}{1000}$ that was used in this chapter is relatively high. (Note that this termination criterion was applied to the change of membership degrees and not prototypes.) By decreasing the precision, some computation time can be saved, because less iteration steps are needed until convergence. A distinct loss in the quality of the analysis results for a termination precision of $\frac{1}{100}$ has not been observed.

Chapter 3

Linear and Ellipsoidal Prototypes

In this chapter, we introduce fuzzy clustering techniques for recognizing linear dependencies between the data. In contrast to conventional regression analysis, however, several clusters can represent more complex piecewise linear relations which can, for instance, be used to construct a fuzzy rule base for function approximation (cf. chapter 8).

In order to illustrate how these algorithms work, we consider two-dimensional examples from image processing. Contiguous areas in images can be reduced to their border lines by so-called *contour operators* well known in image recognition. Our intention is the recognition (resp. approximation) of these border lines using fuzzy clustering.

3.1 The fuzzy c-varieties algorithm

The fuzzy c-varieties algorithm by by H.H. Bock [17] and J.C. Bezdek [10, 11] was developed for the recognition of lines, planes or hyper-planes. Each cluster represents an r-dimensional variety, $r \in \{0, ..., p-1\}$, where p is the dimension of the data space. Each cluster k_i, representing a hyperplane, is characterized by a point v_i and the orthogonal unit vectors $e_{i,1}, e_{i,2}, ..., e_{i,r}$. A cluster is an affine subspace

$$v_i + <e_{i,1}, e_{i,2}, ..., e_{i,r}> = \{y \in \mathbb{R}^p \,|\, y = v_i + \sum_{j=1}^{r} t_j e_{i,j}, \, t \in \mathbb{R}^r\}.$$

For $r = 1$, the variety is a straight line; for $r = 2$, it is a plane; and for $r = p - 1$, it is a hyperplane. In the case $r = 0$, the algorithm is reduced to the recognition of point-shaped clusters and degenerates to the fuzzy c-means. ("Point-shaped" refers to the set of points that have the distance 0 to the cluster.) Here, r is the same for all clusters. The algorithm does not determine r for each cluster individually. The corresponding distance measure is

$$d^2(x, (v, (e_1, e_2, ..., e_r))) \;=\; ||x - v||^2 - \sum_{j=1}^{r}((x - v)^\top e_j)^2.$$

This is the (quadratic) Euclidean distance between a vector x and the r-dimensional variety. For $r = 0$, the sum disappears such that the FCV distance function is identical to the FCM distance function in this case. Given the membership functions, the clusters are obtained according to theorem 3.3. Before we can prove this theorem, we need two short remarks.

Remark 3.1 (Invariance w.r.t. an orthonormal basis) *For each orthonormal basis* $E := \{e_1, e_2, \ldots, e_p\}$ *of* $\mathbb{R}^p$ *and each vector* $x \in \mathbb{R}^p$

$$||x||^2 \;=\; \sum_{j=1}^{p}(x^\top e_j)^2$$

holds.

Proof: Let $S := \{s_1, s_2, \ldots, s_p\}$ be the standard basis of $\mathbb{R}^p$, let $E := \{e_1, e_2, \ldots, e_p\}$ be an arbitrary orthonormal basis of $\mathbb{R}^p$. Since E is a basis, we find for each $i \in \{1, 2, ..., p\}$ factors $a_{1,i}, a_{2,i}, ..., a_{p,i}$ with $\sum_{j=1}^{p} a_{j,i} e_j = s_i$. From the normalization ($e_j^\top e_j = 1$) and the orthogonality ($e_j^\top e_i = 0$ for $i \neq j$) of the two bases, we derive: $1 = ||s_i||^2 = s_i^\top s_i = (\sum_{j=1}^{p} a_{j,i} e_j)^\top (\sum_{j=1}^{p} a_{j,i} e_j) = \sum_{j=1}^{p} a_{j,i}^2 e_j^\top e_j = \sum_{j=1}^{p} a_{j,i}^2$. Thus, $||x||^2 = \sum_{i=1}^{p}(x^\top s_i)^2 = \sum_{i=1}^{p}(\sum_{j=1}^{p} a_{j,i} x^\top e_j)^2 = \sum_{i=1}^{p} \sum_{j=1}^{p}(a_{j,i} x^\top e_j)^2 = \sum_{j=1}^{p}(x^\top e_j)^2 \sum_{i=1}^{p} a_{j,i}^2 = \sum_{j=1}^{p}(x^\top e_j)^2$ also holds . ∎

Remark 3.2 (Meaning of the maximum eigenvalue) *Let* $p \in \mathbb{N}$, $C \in \mathbb{R}^{p \times p}$ *be a matrix, whose normalized eigenvectors* $e_1, e_2, \ldots, e_p$ *form an orthonormal basis of* $\mathbb{R}^p$. *Let* $\lambda_1, \lambda_2, \ldots, \lambda_p$ *be the corresponding eigenvalues. Then we have*

$$\max\{x^\top C x \mid x \in \mathbb{R}^p, \, ||x|| = 1\} \;=\; \max\{\lambda_i \mid i \in \mathbb{N}_{\leq p}\}.$$

Proof: The normalized eigenvectors $e_1, e_2, \ldots, e_p$ form an orthonormal basis of $\mathbb{R}^p$. Let $x \in \mathbb{R}^p$ with $\|x\| = 1$. Then, we can find $\alpha_1, \alpha_2, \ldots, \alpha_p \in \mathbb{R}$ such that $\sum_{i=1}^p \alpha_i e_i = x$. The orthogonality and normalization property of the eigenvectors implies $x^\top C x = x^\top C (\sum_{i=1}^p \alpha_i e_i) = x^\top (\sum_{i=1}^p \alpha_i C e_i) = (\sum_{i=1}^p \alpha_i e_i)^\top (\sum_{i=1}^p \alpha_i \lambda_i e_i) = \sum_{i=1}^p \alpha_i^2 \lambda_i$. For the factors α the normalization of x yields: $1 = \|x\|^2 = x^\top x = (\sum_{i=1}^p \alpha_i e_i)^\top (\sum_{i=1}^p \alpha_i e_i) = \sum_{i=1}^p \alpha_i^2$. Since the sum of the α_i^2 is 1, none of the factors α_i^2 can be greater than 1. The linear combination of the eigenvalues with the factors α_i^2 is greatest, when we choose $\alpha_i = 1$ where i is the index of the greatest eigenvalue. Hence, $x = e_i$. ∎

Theorem 3.3 (Prototypes of FCV) *Let $p \in \mathbb{N}$, $r \in \mathbb{N}_{<p}$, $D := \mathbb{R}^p$, $X = \{x_1, x_2, \ldots, x_n\} \subseteq D$, $C := \mathbb{R}^p \times \{e \in (\mathbb{R}^p)^r \mid \|e_i\| = 1, \ e_i^\top e_j = 0 \text{ for } i \neq j \text{ with } i, j \in \mathbb{N}_{\leq r}\}$, $c \in \mathbb{N}$, $R := \mathcal{P}_c(C)$, J corresponding to (1.7) with $m \in \mathbb{R}_{>1}$ and*

$$d^2 : D \times C \to \mathbb{R}, \ (x, (v, e)) \mapsto \|x - v\|^2 - \sum_{l=1}^r ((x - v)^\top e_l)^2.$$

If J is minimized with respect to all probabilistic cluster partitions $X \to F(K)$ with $K = \{k_1, k_2, \ldots, k_c\} \in R$ and given memberships $f(x_j)(k_i) = u_{i,j}$ with $k_i = (v_i, (e_{i,1}, e_{i,2}, \ldots, e_{i,r}))$ by $f : X \to F(K)$, then

$$v_i = \frac{\sum_{j=1}^n u_{i,j}^m x_j}{\sum_{j=1}^n u_{i,j}^m}$$

$e_{i,l}$ *is the* lth *normalized eigenvector of* C_i *for* $l \in \mathbb{N}_{\leq r}$

$$C_i = \frac{\sum_{j=1}^n u_{i,j}^m (x_j - v_i)(x_j - v_i)^\top}{\sum_{j=1}^n u_{i,j}^m}$$

holds where the eigenvectors are enumerated in such a way that the corresponding eigenvalues form a decreasing sequence.

Proof: The probabilistic cluster partition $f : X \to F(K)$ shall minimize the objective function J. Hence, let $r \in \mathbb{N}_{\leq p}$ and $k \in K$ with $k = (v, (e_1, e_2, \ldots, e_r))$. Then we can find an extension of the basis $\{e_{r+1}, e_{r+2}, \ldots, e_p\}$ for $\{e_1, e_2, \ldots, e_r\}$ such that $E := \{e_1, e_2, \ldots, e_p\}$ is an orthonormal basis of $\mathbb{R}^p$. The Euclidean distance is invariant w.r.t. the orthonormal basis according to remark 3.1, and the distance is

$$d^2(x, (v, (e_1, e_2, \ldots, e_r))) = \|x - v\|^2 - \sum_{l=1}^r ((x - v)^\top e_l)^2$$

$$= \sum_{l=1}^{p}((x-v)^{\top}e_l)^2 - \sum_{l=1}^{r}((x-v)^{\top}e_l)^2$$

$$= \sum_{l=r+1}^{p}((x-v)^{\top}e_l)^2.$$

Since $J(f)$ is a minimum, we must have $\frac{\partial}{\partial v}J(f) = 0$. Defining $u_j = f(x_j)(k)$ we obtain for all $\xi \in \mathbb{R}^p$

$$0 = \frac{\partial}{\partial v}\sum_{j=1}^{n}u_j d^2(x_j,(v,(e_1,e_2,...,e_r)))$$

$$= \frac{\partial}{\partial v}\sum_{j=1}^{n}u_j\sum_{l=r+1}^{p}((x_j-v)^{\top}e_l)^2$$

$$= \sum_{j=1}^{n}u_j\sum_{l=r+1}^{p}\frac{\partial}{\partial v}((x_j-v)^{\top}e_l)^2$$

$$= \sum_{j=1}^{n}u_j\sum_{l=r+1}^{p}-2(x_j-v)^{\top}e_l\cdot\xi^{\top}e_l$$

$$= \left(\sum_{l=r+1}^{p}\left(\sum_{j=1}^{n}u_j(x_j-v)^{\top}e_l\right)e_l^{\top}\right)\xi.$$

Since this equation holds for all ξ, the left expression of the last equation must be identical to 0. However, that expression is a linear combination of $p-r-1$ orthonormal basis vectors so that they can only form a trivial zero sum. Hence, it follows for each $l \in \{r+1, r+2, ..., p\}$

$$0 = \sum_{j=1}^{n}u_j(x_j-v)^{\top}e_l$$

$$\Leftrightarrow \quad \sum_{j=1}^{n}u_j x_j^{\top}e_l = \sum_{j=1}^{n}u_j v^{\top}e_l$$

$$\Leftrightarrow \quad v^{\top}e_l = \frac{\sum_{j=1}^{n}u_j x_j^{\top}e_l}{\sum_{j=1}^{n}u_j}.$$

The last equation states that the lth coordinate of v in the E-system results from the lth coordinates of all x_j in the E-system. If we introduce a coordinate transformation $\psi : \mathbb{R}^p \to \mathbb{R}^p$ which transforms the vectors of $\mathbb{R}^p$ from the standard coordinates to E coordinates, the equation becomes

$$\psi(v)_l = \frac{\sum_{j=1}^{n}u_j\psi(x_j)_l}{\sum_{j=1}^{n}u_j}, \tag{3.1}$$

where y_l denotes the lth coordinate of a vector y. The coordinates $r+1, r+2, ..., p$ are thus given. Since the first r basis vectors determine our variety (up to a translation), we can assume arbitrary coordinates for them, because we always remain within the variety, i.e. we do not change the distance between the datum and the variety. Hence, we choose those coordinates analogously to (3.1). Thus, v can be written as

$$\psi(v) = \frac{\sum_{j=1}^{n} u_j \psi(x_j)}{\sum_{j=1}^{n} u_j}. \tag{3.2}$$

Since the mapping ψ as a linear mapping with $\psi(\mathbb{R}^p) = \mathbb{R}^p$ is a vector space isomorphism (and hence also ψ^{-1}), it follows from (3.2):

$$
\begin{aligned}
v &= \psi^{-1}(\psi(v)) \\
&= \psi^{-1}\left(\frac{\sum_{j=1}^{n} u_j \psi(x_j)}{\sum_{j=1}^{n} u_j}\right) \\
&= \frac{\sum_{j=1}^{n} u_j \psi^{-1}(\psi(x_j))}{\sum_{j=1}^{n} u_j} \\
&= \frac{\sum_{j=1}^{n} u_j x_j}{\sum_{j=1}^{n} u_j} \quad .
\end{aligned}
$$

We still have to determine the direction vectors that induce the variety. Without loss of generality we assume that the zero vector is the position vector for the variety. Since the direction vectors appear in the objective function only in non-positive expressions, the objective function is smallest at f, when $\sum_{j=1}^{n} u_j (x_j^\top e_l)^2 = \sum_{j=1}^{n} u_j e_l^\top (x_j x_j^\top) e_l = e_l^\top \left(\sum_{j=1}^{n} u_j x_j x_j^\top\right) e_l$ is maximum for each $l \in \mathbb{N}_{\leq r}$. From remark 3.2 we derive that the e_l, $l \in \mathbb{N}_{\leq r}$ are the r eigenvalues of the matrix $\sum_{j=1}^{n} u_j x_j x_j^\top$ corresponding to the greatest eigenvalues. ∎

Figures 3.1 and 3.2 illustrate how the fuzzy c-varieties algorithm can recognize lines. The three intersecting lines were recognized, as well as the four boundaries of the twisted rectangle. In both figures, however, the areas of high memberships exceed beyond the line segments. In image processing, one is interested not so much in straight lines but rather in straight line segments as boundaries of objects. However, a one-dimensional variety, a straight line, always extends to infinity. In figures 3.1 and 3.2, there are no data in the further course of the straight line so that the infinite expansion

Figure 3.1: FCV analysis　　　　Figure 3.2: FCV analysis

of the clusters does not lead to any undesired side effects. However, this is completely different in figure 3.3.

Each pair of the four straight line segments in this example lies on a straight line. Since each cluster expands to infinity, the algorithm recognizes each pair of straight line segments correctly as one straight line. For an optimum partition in the sense of a minimization of the objective function, two clusters suffice for FCV. The optimum partition in the sense of image processing, however, aims at straight line segments, i.e. a partition into four clusters. Four clusters are too many for FCV here. The memberships to the additional clusters are therefore very small. Only at the intersections between the original clusters do they have higher memberships. In the extreme case, only two data (one on each intersection point) are assigned to the superfluous clusters this way. These few high memberships stabilize the cluster in that (arbitrary) orientation, since no further data effect the orientation of the clusters.

In the previous example, the initialization of the prototypes clearly influenced the clustering result. By default, we initialized the positions of the prototypes as equidistant points along the x-axis. With this initialization the algorithm recognized four nearly parallel, vertical lines (figure 3.3). The initialization is close to a local minimum. Hence, a fuzzy c-means was run as an initialization for five steps. This already moves the prototypes close to their intuitively correct position. Nevertheless, the collinear line segments are merged together again because of the infinite expansion of the straight line clusters.

Figure 3.3: FCV analysis

A higher number of clusters increases the number of local minima (figure 3.4). If the data are distributed over a few clusters in the course of the algorithm, all remaining data can be covered by the remaining clusters because of their infinite extension, even when they are very far away.

In figure 3.4, the long edges of the lower rectangle, with their high numbers of pixels, cause a fast orientation of all clusters in that direction. A majority of the data is now covered by only two clusters. The remaining clusters cover all the rest of the data because of their relatively high number and parallel orientation. Even if this result is (locally) optimal in the sense of the objective function, it is useless for image recognition. For image processing, the algorithm for the recognition of straight line segments is only partly useful.

A possible solution for this problem was suggested by Bezdek [10], and it is based on a change in the distance function. In order to consider also the extension of the data within the affine subspace, the Euclidean distance between the data and the position vector v of the prototype (v, e) is added in a certain proportion to the distance function in theorem 3.3:

$$d^2_{FCE}(x, (v, e)) = \alpha d^2(x, (v, e)) + (1 - \alpha)||x - v||^2.$$

When choosing the prototypes corresponding to theorem 3.3, the resulting *fuzzy c-elliptotypes algorithm* does a better job in the sense of the recognition of straight line segments. Because of the factor α, a deformation of the clusters from spherical ($\alpha = 0$) to linear ($\alpha = 1$) can be achieved.

Figure 3.4: FCV analysis

For the shapes between the extremes, Bezdek created the notion *elliptotype*. With $\alpha = 0.9$, some of the FCV problem cases can already be clustered successfully. The α value has yet to be chosen once for all clusters. (See also section 3.2.)

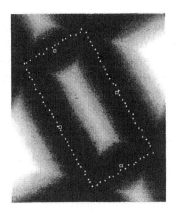
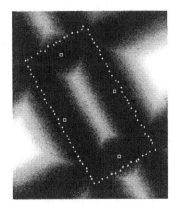

Figure 3.5: $i = 0$ Figure 3.6: $i = 10$

A further problem arises for the fuzzy c-varieties algorithm when a possibilistic cluster partition is determined. Along a straight line, the distances to the data vectors are very low, and the memberships in these straight lines are very high. Because of the normalization of the memberships for *probabilistic* clustering, the memberships of these data in the other clusters

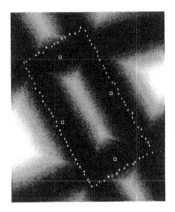

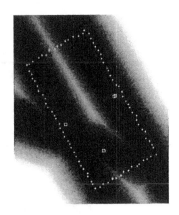

Figure 3.7: $i = 20$ Figure 3.8: $i = 40$

become very low. This effect no longer occurs with possibilistic clustering. In the first step of possiblistic clustering, the two clusters of the shorter straight line segments from figure 3.5 already move a little bit in the direction of the rectangle's centre. The reason is that the points of the longer lines near the short edges now get a higher membership in the clusters of the short edges. The comparatively much lower distance to the clusters of the longer edges is not taken into account when the memberships are assigned. This way, all the clusters move in the direction of the rectangle's centre. Here, the shorter edges can make further moves since the clusters across the longer edges are kept in their position by fewer data. Figure 3.6 shows the state after ten steps. Because of the irregular distribution of the data on the rectangle's borders, it may happen that a straight line is not shifted parallel to the original straight line. This tendency can already be observed for the upper cluster after only 20 iteration steps, as shown in figure 3.7. After 40 iteration steps (cf. figure 3.8), the final result can be easily predicted: both clusters of the shorter rectangle's lines move away fast and turn to the orientation of the longer lines. The (formerly) upper cluster already coincides with the right cluster; the cluster partition is unsatisfactory. Again in such cases noise clustering outperforms possibilistic clustering, since membership normalization prevents well detected clusters from influencing themselves.

3.2   The adaptive fuzzy clustering algorithm (fuzzy c-elliptotypes)

In order to be able to recognize straight line segments instead of straight lines, the adaptive fuzzy clustering algorithm was developed. By considering the distance between points in the line segments and the elliptotype "centre" v, disjoint line segments are assigned to different clusters. The modified distance function of the fuzzy c-varieties is used, as already mentioned in the previous section. With the denotations from theorem 3.3, we define

$$d^2 : D \times C \to \mathbb{R},$$

$$
\begin{aligned}
(x, (v, e)) \quad \mapsto \quad & \alpha \left(||x - v||^2 - \sum_{j=1}^{r} ((x - v)^\top e_j)^2 \right) + (1 - \alpha) ||x - v||^2 \\
= \quad & ||x - v||^2 - \alpha \sum_{j=1}^{r} ((x - v)^\top e_j)^2.
\end{aligned}
$$

By the choice of α, the shape can be changed from point-shaped clusters ($\alpha = 0$) via elliptic shapes ($\alpha \in \,]0, 1[$) to straight lines ($\alpha = 1$). With an α whose value is constant for all clusters, this modification of the fuzzy c-varieties is known as fuzzy c-elliptotypes [10]. Davé made a suggestion for the choice of α for each single cluster [22]. We use the denotations from theorem 3.3 for cluster k_i. If the eigenvalues of the matrix C_i are $\lambda_{i,1}, \lambda_{i,2}, ..., \lambda_{i,p}$ in descending order, choose for $i \in \mathbb{N}_{\leq c}$:

$$\alpha_i = 1 - \frac{\lambda_{i,p}}{\lambda_{i,1}}.$$

The prototypes k_i are determined in the same way as in theorem 3.3.

This heuristic improvement works well for the two-dimensional case only. The eigenvalues provide information about the expansion of the fuzzy scatter matrix into the direction of the eigenvectors. For a linear cluster, one of the expansions is equal to 0, so that α becomes equal to one. In this case, the cluster behaves as it does with the fuzzy c-varieties algorithm. If we do not deal with an ideal straight line, the proportion between the shorter and the longer expansion defines the value α. The more similar the eigenvalues are, the more spherical is the cluster. In this case, α gets close to zero. The cluster thus behaves as it does with the fuzzy c-means.

For straight lines in the three-dimensional space this approach does not work any longer. If the data vectors are located exactly in a plane, the

Figure 3.9: AFC analysis   Figure 3.10: AFC analysis

expansion vertical to this plane is 0. Thus, one eigenvalue will be 0. Hence, α becomes 1, and the algorithm searches for ideal straight lines, although the cluster may have a circular expansion within the plane.

Figure 3.11: AFC analysis

Since image processing deals with points in $\mathbb{R}^2$, Davé's approach is suitable for such applications. For instance, the cuboid in figure 3.9 is correctly recognized by the adaptive fuzzy clustering algorithm. Also, for the data set from figure 3.4 with two intersecting rectangles, complicating the correct separation of the data, the result corresponds to the intuitive partition, as shown in figure 3.10.

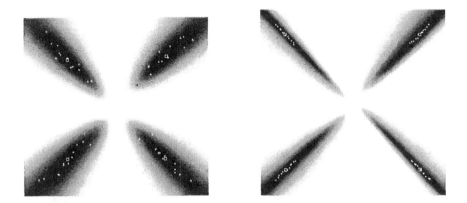

Figure 3.12: AFC analysis   Figure 3.13: AFC analysis

However, the algorithm does not completely solve the problem indicated in figure 3.3. Figures 3.11, 3.12 and 3.13 show similar data sets. Here, only the latter two figures were correctly recognized, while for the first figure we see the problems known from the fuzzy c-varieties algorithm: the straight line segments (bottom left and top right) are interpreted as one cluster. (Notice the small square near the middle between the two clusters which marks the position vector of the cluster. If the corresponding cluster would approximate only one of the straight line segments, the vector would be near the centre of the respective straight line segment.) The adaptive fuzzy clustering algorithm tries to adapt the expansion of linear clusters by choosing individual parameters α for each cluster.

The data set from figure 3.12 shows a larger scattering of the data along the straight line. Thus α is decreased and the algorithm searches for shorter, elliptic clusters. The scattering directly influences the α values, and thus the cluster partition.

Also, by pulling the clusters apart, as can be seen in figure 3.13, the originally bad partition of the data set from figure 3.11 can be improved. The α factors control the influence of the Euclidean distance between the data and the centre of the variety, so that different partitions can be achieved by moving the clusters within the variety. With usual α values close to 1, the distances between clusters must be distinctly greater than the distances of the data vectors to the straight line in order to influence the separation of the clusters.

In figure 3.11, the cluster distances for the determined α values are too low to cause a separation of the clusters; they are sufficient in figure 3.13.

Especially for the data sets just considered, good results can be obtained with an initialization by fuzzy c-means, because for these data sets, there is already an almost correct partition of the data after FCM. For intersecting lines, however, a fuzzy c-means initialization does not help; the problems described above are likely to occur again in a different form.

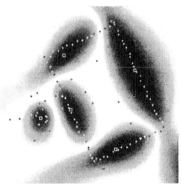
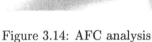
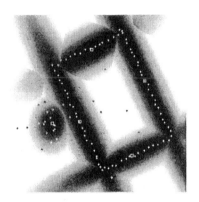

Figure 3.14: AFC analysis   Figure 3.15: P-AFC analysis

All the previous examples tried to recognize lines or line segments, but the algorithm is also able to recognize elliptic or circular clusters. A data set with four low noise line segments and a point cloud is shown in figure 3.14. The algorithm recognizes the shapes well. Possibilistic clustering is also feasible with the adaptive fuzzy clustering algorithm, as shown in figure 3.15. With the FCV algorithm the influence of the bordering rectangle's edges caused a degeneration of the partition (cf. figure 3.5). By additionally considering the Euclidean distance from the position vector, these edges obtain larger distance values which prevents the degeneration.

If the point cloud of the previous example is shifted down along the left edge of the rectangle, it gets into the influence of the lower linear cluster. Since the memberships of the lower cluster are very high even beyond the line segment, this modification moves the linear cluster to the point cloud. This effect cannot be compensated for even by possibilistic clustering. The expansion of the clusters is not infinite with the adaptive fuzzy clustering algorithm, but it is still very large for clusters induced by data on straight lines. In this way, the problems known from the fuzzy c-varieties might still occur. Intuitively, the memberships should decrease noticeably faster at the end of the straight line segment than it is the case, even with the adaptive fuzzy clustering algorithm.

From the perspective of the implementation, hardly anything changes compared to the fuzzy c-varieties algorithm, since the eigenvalues have to be computed to update the prototypes anyway. The additional time for determinating the values for α can almost be neglected.

3.3   The Gustafson-Kessel algorithm and the Gath-Geva algorithm

The Gustafson-Kessel and the Gath-Geva algorithms were already introduced in sections 2.2 and 2.3. The example data of the corresponding sections contain lines, too, which were recognized well by the algorithms. Therefore, these algorithms shall be tested once again in this section with respect to the recognition of lines, in order to compare them to the fuzzy c-varieties or adaptive fuzzy clustering algorithm. All examples in this section were exclusively clustered by the Gustafson-Kessel algorithm. Similar results were obtained by the Gath-Geva algorithm; for a comparison of the algorithms, see section 2.3.

Figure 3.16: GK analysis

Figure 3.17: GK analysis

As shown in figures 3.16 and 3.17, the Gustafson-Kessel algorithm is also able to partition those data correctly which cause problems for the fuzzy c-varieties algorithm. Because of the high effort for the determination of the eigenvalues and eigenvectors, the Gustafson-Kessel algorithm is even faster than the adaptive fuzzy clustering algorithm. (Figure 3.17 was correctly clustered by the AFC too, without showing the result here.)

For the data set from figure 3.14, the Gustafson-Kessel algorithm also provides a good result. Comparing the possibilistic runs (figures 3.15 and 3.19), a different behaviour of the two algorithms can be clearly recognized. The Gustafson-Kessel clusters seem to be narrower and the areas

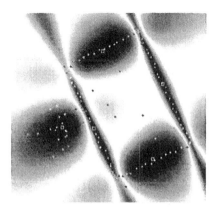

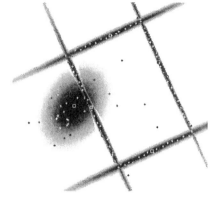

Figure 3.18: GK analysis

Figure 3.19: P-GK analysis

with higher memberships are thinner. The point cloud is very differently clustered by the two algorithms. With the Gustafson-Kessel algorithm, the extremely stretched ellipses for the straight line clusters deform the distance vertical to the straight line so strongly that even relatively close points obtain only low memberships. These are then assigned to the elliptical shaped cluster. With the adaptive fuzzy clustering algorithm this effect does not occur. There, these points are correctly assigned to the linear cluster.

For the three data sets from figures 3.11, 3.12 and 3.13, the Gustafson-Kessel algorithm proved to be superior. With four clusters to be recognized, the resulting partition corresponded to the intuitive expectation each time. The variation of the scattering or distance had no negative influence on the partition. Instead, for $c = 2$ the analysis results were worse than they were with the adaptive fuzzy clustering algorithm; the left and right clusters, respectively, were merged.

The Gustafson-Kessel (and the Gath-Geva) algorithm does not require additional modifications to detect lines. For some examples, the *classical* algorithms seem to be even superior for the recognition of straight line segments.

3.4 Computational effort

The data sets in this section contained between 50 and 200 data. With the simple FCV examples, the algorithm needed only a few (≈ 10) steps, in the difficult case 3.4 it took 87. The analyses by the AFC algorithm were computed within 20 to 50 steps, the GK algorithm required an average of 50 steps and a maximum of 184 steps for the example shown in figure 3.17.

The comparison of the time indices for the three linear clustering techniques shows only small differences – at least in the two-dimensional case: $\tau_{FCV} = 1.46 \cdot 10^{-4}$, $\tau_{AFC} = 1.48 \cdot 10^{-4}$ and $\tau_{GK} = 1.47 \cdot 10^{-4}$. The inversion of a 2×2-matrix took as long as the search for the eigenvalues and eigenvectors of the matrix. However, this may be different for higher dimensions.

Chapter 4

Shell Prototypes

While there are many different application areas for the solid clustering techniques from chapter 2, and the linear clustering techniques from chapter 3, we discuss the rather special shell clustering techniques in this chapter. Instead of elliptic clusters, we are now looking for object boundaries, shells or edges. Of course, there might be applications where measured data are on the shell of an hyper-ellipsoid whose parameters can be determined with these techniques. However, the main application field of these techniques lies in image processing, mainly because of their computational infeasibility in higher dimensions.

Before those techniques can be applied to real images, though preprocessing is necessary. Up to now, we have always assumed crisp data sets that contain the coordinates of all the dark pixels, in the case of image recognition. If we want to determine the parameters of a circle-shaped plate in a picture, we have to preprocess the image data so that the contours of the plate remain in the data set. A great variety of so-called contour operators was developed in image recognition. A contour operator determines the pixels' new intensities from the changes in brightness. If we are in the middle of a homogeneous area of one colour, there are no changes in brightness; the pixel will get a low intensity. We recognize an edge of an object by the different brightness values to the left and to the right of an edge; depending on the contrast, we obtain a high intensity for the pixels this way. We now include all points in the data set for the fuzzy cluster analysis which have a certain minimum intensity after the contour processing. In this way, information about the unambiguity of the mapped edges is lost, of course, because we can not distinguish the intensities of the data after the application of a threshold any longer. On the other hand, pixels in a real image usually have different brightness values; instead, there are even certain variations in brightness in homogeneous areas, too, because of object structures or reflections. The contour operator correspondingly produces small fluctuations in the intensity for almost all pixels. If we want to use all these data as grey values in the analysis, our data set will consist of several hundred thousand data which can not be handled any more by the introduced methods in a feasible way. The application of a thresholding technique to reduce the data to a few thousand pixels can be recommended. We should not necessarily ignore the information about the intensity, though, i.e. the quality of the represented contour. If the intensity is available as a weight $w_j \in [0, 1]$ for each $x_j \in X$, the w_j can be included in the objective function as an additional factor: $\sum_{i=1}^{c} \sum_{j=1}^{n} w_j u_{i,j}^m d_{i,j}^2$. These factors can be kept for the derivations of the prototypes leading to an easy analysis of grey value.

4.1   The fuzzy c-shells algorithm

The first algorithm for the recognition of circle contours (fuzzy c-shells, FCS) was introduced by Davé [23]. Each cluster is characterized by its centre v and radius r. The (Euclidean) distance of a datum x to a circle (v, r) is $|\, \|x - v\| - r\,|$. This leads to

Theorem 4.1 (Prototypes of FCS) *Let* $p \in \mathbb{N}$, $D := \mathbb{R}^p$, $X = \{x_1, x_2, \ldots, x_n\} \subseteq D$, $C := \mathbb{R}^p \times \mathbb{R}_{>0}$, $c \in \mathbb{N}$, $R := \mathcal{P}_c(C)$, J *corresponding to* (1.7) *with* $m \in \mathbb{R}_{>1}$ *and*

$$d^2 : D \times C \to \mathbb{R}, \ (x, (v, r)) \mapsto (\|x - v\| - r)^2.$$

If J *is minimized with respect to all probabilistic cluster partitions* $X \to F(K)$ *with* $K = \{k_1, k_2, \ldots, k_c\} \in R$ *and given memberships* $f(x_j)(k_i) = u_{i,j}$ *with* $k_i = (v_i, A_i)$ *by* $f : X \to F(K)$, *then*

$$0 = \sum_{j=1}^{n} u_{i,j}^m \left(1 - \frac{r_i}{\|x_j - v_i\|} \right) (x_j - v_i) \qquad (4.1)$$

$$0 = \sum_{j=1}^{n} u_{i,j}^m \left(\|x_j - v_i\| - r_i \right) \qquad (4.2)$$

holds.

   Proof: The probabilistic cluster partition $f : X \to F(K)$ shall minimize the objective function J. Let $i \in \mathbb{N}_{\leq c}$ and $k_i = (v_i, r_i) \in K$. Then, the partial derivative of J with respect to r_i and the directional derivative of J with respect to v_i are necessarily 0. Thus, we have for all $\xi \in \mathbb{R}^p$:

$$
\begin{aligned}
0 &= \frac{\partial}{\partial v_i} J \\
&= \sum_{j=1}^{n} u_{i,j}^m \frac{\partial}{\partial v_i} (\|x_j - v_i\| - r_i)^2 \\
&= 2 \sum_{j=1}^{n} u_{i,j}^m (\|x_j - v_i\| - r_i) \frac{\partial}{\partial v_i} \sqrt{(x_j - v_i)^\top (x_j - v_i)} \\
&= 2 \sum_{j=1}^{n} u_{i,j}^m (\|x_j - v_i\| - r_i) \frac{-\xi^\top (x_j - v_i) - (x_j - v_i)^\top \xi}{2\sqrt{(x_j - v_i)^\top (x_j - v_i)}} \\
&= -2 \sum_{j=1}^{n} u_{i,j}^m (\|x_j - v_i\| - r_i) \frac{1}{\|x_j - v_i\|} (x_j - v_i)^\top \xi \\
&= -2 \left(\sum_{j=1}^{n} u_{i,j}^m \left(1 - \frac{r_i}{\|x_j - v_i\|} \right) (x_j - v_i)^\top \right) \xi.
\end{aligned}
$$

Because the last expression vanishes for all $\xi \in \mathbb{R}^p$, we can write instead

$$
0 = \sum_{j=1}^{n} u_{i,j}^m \left(1 - \frac{r_i}{\|x_j - v_i\|} \right) (x_j - v_i).
$$

Taking the derivative with respect to r_i we obtain

$$
\begin{aligned}
0 &= \frac{\partial}{\partial r_i} J \\
&= \sum_{j=1}^{n} u_{i,j}^m \frac{\partial}{\partial r_i} (\|x_j - v_i\| - r_i)^2 \\
&= -2 \sum_{j=1}^{n} u_{i,j}^m (\|x_j - v_i\| - r_i) \\
\Leftrightarrow \quad 0 &= \sum_{j=1}^{n} u_{i,j}^m (\|x_j - v_i\| - r_i).
\end{aligned}
$$

Unfortunately, theorem 4.1 does not provide an explicit solution for the determination of the prototypes. The equations (4.1) and (4.2) form a non-linear, coupled, $(p+1)$-dimensional system of equations for $p+1$ variables. It can be solved, for instance, by the iterative Newton algorithm. The result of the previous FCS step can be used as the initialization for the following Newton iteration. For the first iteration step of the fuzzy c-shells algorithm, the centres v_i can be initialized by the fuzzy c-means algorithm, and the radii for $i \in \mathbb{N}_{\leq c}$ by

$$r_i = \frac{\sum_{j=1}^{n} u_{i,j}^m \|x_j - v_i\|}{\sum_{j=1}^{n} u_{i,j}^m}. \tag{4.3}$$

This expression determines the mean distance of the centre point to the associated data.

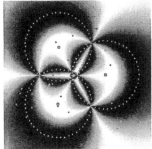

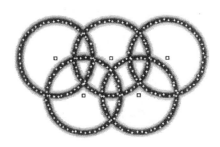

Figure 4.1: FCS analysis   Figure 4.2: FCS analysis

The three circles in figure 4.1 are recognized very well by the probabilistic algorithm, despite of 30 noise data items. Also, the data of the five interlaced circles from figure 4.2 are clearly separated. By the way, for this figure a different kind of representation was chosen. The usual representation of the memberships as in figure 4.1 is getting extremely confusing with the increasing number of clusters. In figure 4.2, the *probabilistic* partition was represented *possibilistically* by simply choosing all extension factors as constanst. We will use that kind of representation for better clarity in all of the following chapters.

Even if the data do not form a complete circle, the fuzzy c-shells algorithm recognizes the circle segments well, as in the lower left cluster in figure 4.3. For the upper left cluster, the circle has certainly small distances to the associated data, but it does not fit into the intuitive partition. The reason is that the circle's segment is very small. The cluster had a completely different position in the beginning, but there always remain some data at the intersection points with the other circles that keep their high memberships. Since it is easy to fit a circle to the remaining data of the old cluster shape and the data of the short circle segment so that all data are approximated at the same time, we obtain an undesired cluster as in figure 4.3. There is a similar phenomenon for the linear clustering algorithms that show the tendency to group collinear data – regardless of their Euclidean distance – in one cluster. Shell clustering does not necessarily require data which are close together. Theoretically, any three arbitrary point accumulations can be minimized by a single circle-shaped cluster. In this way, the number of partitions corresponding to local minima increases considerably.

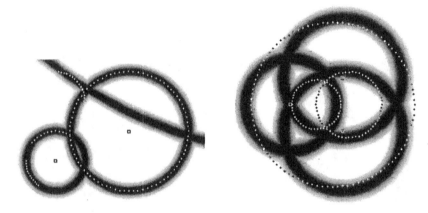

Figure 4.3: FCS analysis   Figure 4.4: FCS analysis

With the data from figure 4.4, the algorithm got stuck in a local minimum. Parts of different circles were united to one cluster. The tendency to converge in a local minimum is distinctly greater for mutually enclosing circles than for the relatively small overlaps in the previous examples. If one circle lies within another circle, it will hardly leave it in the course of the iteration, since all outside points have a smaller distance to the enclosing circle. The same problem occurs, when two initially separated circles are supposed to enclose each other afterwards. Whether the final clustering result meets the intuition depends strongly on the initialization.

As already mentioned in chapter 1, high values for the fuzzifier m cause the objective function to be smoother so that the influence of local minima is reduced or they even disappear completely. Therefore, $m > 2$ should yield better results. For instance, an FCS run with $m = 8.0$ after a fuzzy c-means initialization with $m = 10.0$ does in fact recognize all circles correctly. (However, an enlarged fuzzifier does not lead to a better partition for all examples.) In the literature [72], it is proposed to initialize circle algorithms for enclosing circles, in such a way that the circles' centre points are scattered around the centre point of all data. This is exactly what we achieved by choosing a high fuzzifier for FCM.

The fuzzy c-shells algorithm is computationally expensive because of the Newton iteration within the clustering iteration. In the implementation of the algorithm that was used here, a modified Newton procedure according to [32] was applied. This *iteration in the iteration* mainly influences the computing time because the single steps of the Newton procedure are computationally intensive themselves – for example, the inverted Jacobian matrix has to be computed in (almost) every step. Therefore, it is recommended to optimize each cluster by a Newton iteration of its own with $p + 1$ equations, instead of solving the system of $c \cdot (p + 1)$ equations in a single iteration. In the case of image recognition, the computations with a dimension of $p + 1 = 3$ remain acceptable.

Bezdek and Hathaway showed [12] that it is not necessary to solve the equation system in each step exactly. This way, the time needed for each Newton iteration is decreased, but the number of fuzzy c-shells iterations is possibly increased. In the implementation we applied here, the precision for termination was chosen dependent on the progress of the fuzzy c-shells iteration. It is pointless to compute the cluster parameters very precisely as long as the memberships still vary to a large extent. If they become more stable, though, we are closer to the final clustering result, and the partition of the clusters should be carried out as exactly as possible. When δ is the maximal alteration of memberships compared to the last (fuzzy c-shells) iteration step and ε is the (fuzzy c-shells) precision to be achieved, then the Newton iteration was carried out until the precision $\max\{\varepsilon, \frac{\delta}{10}\}$ was reached. A significant increase in the number of fuzzy c-shells iteration steps could not be observed here. For the application of this algorithm, it has to be taken into account that *at least one iteration step is carried out* with the Newton algorithm, even if the initial value in the norm is already below the termination criterion. If this step is skipped, the prototypes remain unchanged, and thus the memberships too, if the current (Newton) termination precision is greater than the previously reached precision. Since there is no alteration of the memberships, the program then consequently stops because it is deemed to have reached the termination precision ε, although the resulting partition is not minimum at all.

Furthermore, it should be taken into account for the implementation that the n singularities of equation (4.1) do not lead to a irregular termination of the program. Of course, it may happen that a datum is exactly located on a circle's centre, and that leads to a division by zero. This datum, which is responsible for the error, is certainly not an especially typical

datum of the circle on whose centre it is lying. If we set a value close to the
machine precision for the distance instead of 0 in that case, we avoid an
error without distorting the overall result of the fuzzy c-shells within the
computation precision.

4.2 The fuzzy c-spherical shells algorithm

The largest drawback of the fuzzy c-shells algorithm is its high computational complexity because of the implicit equations for the prototypes in
theorem 4.1. Using a different distance measure, the so-called algebraic
distance, prototypes can be given explicitly [83, 67]. Thus, the necessary
computing time is significantly reduced.

Theorem 4.2 (Prototypes of FCSS) *Let* $p \in \mathbb{N}$, $D := \mathbb{R}^p$, $X = \{x_1, x_2, \ldots, x_n\} \subseteq D$, $C := \mathbb{R}^p \times \mathbb{R}_{>0}$, $c \in \mathbb{N}$, $R := \mathcal{P}_c(C)$, J *corresponding to* (1.7) *with* $m \in \mathbb{R}_{>1}$ *and*

$$d^2 : D \times C \to \mathbb{R}, \ (x, (v, r)) \mapsto (\|x - v\|^2 - r^2)^2.$$

If J *is minimized with respect to all probabilistic cluster partitions* $X \to F(K)$ *with* $K = \{k_1, k_2, \ldots, k_c\} \in R$ *and given memberships* $f(x_j)(k_i) = u_{i,j}$ *with* $k_i = (v_i, r_i)$ *by* $f : X \to F(K)$, *then*

$$v_i = -\frac{1}{2}(q_{i,1}, q_{i,2}, \ldots, q_{i,p})^\top$$

$$r_i = \sqrt{v_i^\top v_i - q_{i,p+1}}$$

and

$$q_i = -\frac{1}{2}H_i^{-1}w_i$$

$$H_i = \sum_{j=1}^{n} u_{i,j}^m \, y_j y_j^\top$$

$$w_i = \sum_{j=1}^{n} u_{i,j}^m s_j$$

$$s_j = 2(x_j^\top x_j)y_j \quad \text{for all } j \in \mathbb{N}_{\leq n}$$

$$y_j = (x_{j,1}, x_{j,2}, \ldots, x_{j,p}, 1)^\top \quad \text{for all } j \in \mathbb{N}_{\leq n}$$

holds.

Proof: The probabilistic cluster partition $f : X \to F(K)$ shall minimize the objective function J. The algebraic distance $d^2(x,(v,r)) = (\|x - v\|^2 - r^2)^2$ can be written as $d^2(x,(v,r)) = q^\top M q + s^\top q + b$ where $q = (-2v_1, -2v_2, ..., -2v_p, v^\top v - r^2)$, $M = yy^\top$, $y = (x_1, x_2, ..., x_p, 1)$, $s = 2(x^\top x)y$ and $b = (x^\top x)^2$. All unknown variables are exclusively contained in q. If we interpret J as a function of q_i instead of (v_i, r_i), we derive from the minimality of J at f that $\frac{\partial}{\partial q_i} b = 0$ must hold for all $i \in \mathbb{N}_{\leq c}$. Thus, we have for all $\xi \in \mathbb{R}^{p+1}$:

$$
\begin{aligned}
0 &= \frac{\partial}{\partial q_i} J \\
&= \sum_{j=1}^{n} u_{i,j}^m \frac{\partial}{\partial q_i} d^2(x_j, q_i) \\
&= \sum_{j=1}^{n} u_{i,j}^m (\xi^\top M_j q_i + q_i^\top M_j \xi + \xi^\top s_j) \\
&= \sum_{j=1}^{n} u_{i,j}^m (\xi^\top M_j q_i + (M_j q_i)^\top \xi + \xi^\top s_j) \quad (M_j \text{ symmetric}) \\
&= \sum_{j=1}^{n} u_{i,j}^m (\xi^\top M_j q_i + (\xi^\top M_j q_i)^\top + \xi^\top s_j) \quad (\xi^\top M_j q_i \in \mathbb{R}^{1 \times 1}) \\
&= \xi^\top \left(\sum_{j=1}^{n} u_{i,j}^m (2 M_j q_i + s_j) \right).
\end{aligned}
$$

Since the last expression disappears for all $\xi \in \mathbb{R}^{p+1}$, it follows that

$$
\begin{aligned}
0 &= \sum_{j=1}^n u_{i,j}^m (2 M_j q_i + s_j) \\
\Leftrightarrow \quad -\sum_{j=1}^n u_{i,j}^m s_j &= 2 \left(\sum_{j=1}^n u_{i,j}^m M_j \right) q_i \\
\Leftrightarrow \quad q_i &= -\tfrac{1}{2} \left(\sum_{j=1}^n u_{i,j}^m M_j \right)^{-1} \left(\sum_{j=1}^n u_{i,j}^m s_j \right) \\
&= -\tfrac{1}{2} H_i^{-1} w_i .
\end{aligned}
$$

The matrix H_i is invertible, if the data set X contains $(p+1)$ linearly independent data. Knowing q, we can directly compute the centre and radius as stated in the theorem. ∎

Thus, there are two algorithms for the recognition of circles, which differ in their distance function. For the fuzzy c-spherical shells algorithm, the square of the distance is used. In this way, data inside and outside a circle, which have the same Euclidean distance from the circle line, will be assigned different distances. Considering a circle with radius $\frac{1}{2}$ and centre at $(0,0)$, the datum $(1,0) \in \mathbb{R}^2$ has the distance $(1 - \frac{1}{4})^2 = \frac{9}{16} = 0.5625$, and the datum $(0,0) \in \mathbb{R}^2$ has the distance $(0 - \frac{1}{4})^2 = \frac{1}{16} = 0.0625$. The fuzzy c-spherical shells thus react to the noise data outside the circle more than to the data inside the circle. However, this also means that data which should be almost exactly on the cluster shell, but are within the cluster circle due to distant noise data, can hardly correct the circle's parameters. The distance function of the fuzzy c-shells algorithm, on the other hand, rates such data equally independent of their position.

However, the results of the two algorithms are often similar with regard to their performance in recognition. With one exception, the same partitions were achieved for the data sets of section 4.1. The exception was the data set from figure 4.3, where the upper left cluster was recognized according to intuition. Here, the changed norm had a positive effect.

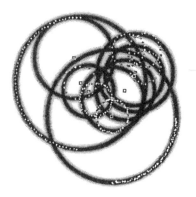

Figure 4.5: FCSS analysis

Figure 4.6: FCSS analysis

Because of the explicit equations for the prototypes, the fuzzy c-spherical shells algorithm provides the clustering result faster than the fuzzy c-shells algorithm. However, this gain in time does not mean that more complicated data sets can be correctly partitioned in the same time now. As with linear clusters, for larger number of clusters there is the risk of a partition that simply covers almost the whole image area, but which does not correspond to the intuitive partition. Such a case is shown in

figure 4.5, where the fuzzy c-spherical shells algorithm was initialized by
only five fuzzy c-means steps. Fifteen further FCM steps result in a distinctly better partition by FCSS already, as shown in figure 4.6. The only
disturbing factor is the coverage of the upper left semicircle by two circle
clusters, because there is now one cluster missing in the area of the three
small circles on the above right. Here, the higher number of FCM steps
has its disadvantages, too. The semicircle was divided into two clusters by
the fuzzy c-means because of its big radius, however, the associated data
describes the same circle. Instead of an approximation of the semicircle
by one cluster, the semicircle and parts of the three small circles above
are now covered by two clusters. Thus, even the number of FCM iteration
steps for the initialization has a great influence on the clustering result.
A possibilistic run causes (in both cases) hardly any changes. The high
memberships for the data near the intersection points of the circle cluster
prevent great changes. Also, an increase of the fuzzifier can hardly achieve
anything, presumably because the extent of the local minima is very strong
according to the non-linear algebraic distance function.

Man and Gath proposed a further algorithm for detecting circles, the
fuzzy c-rings algorithm (FCR) [72]. Although they consider the Euclidean
distance function of the FCS, they do not derive the FCR prototypes in
the way that is usual for the family of fuzzy clustering techniques introduced. For the computation of the prototypes, the old prototypes (from
the previous iteration step) are included. (For example, the new radius is
computed by solving equation (4.2) with respect to r_i, where the old centre
is assumed for v_i.) Thus, the new prototype is not adopted to the data in
the optimal way but just improved. The general convergence considerations
by Bezdek cannot be applied to this algorithm any more. It could at best
be argued that computations should not to be carried out until the exact
minimum is reached when the Newton algorithm is applied, in order to obtain convergence. However, there is already an algorithm, which is easy to
compute, available for the recognition of circles, viz. the FCSS algorithm,
and therefore we do not discuss FCR here in detail. For the recognition
of ellipses, though, we will return to this subject once again when dealing
with the FCE algorithm (cf. section 4.5).

4.3   The adaptive fuzzy c-shells algorithm

The recognition of circles alone is not satisfactory, because the projection
of circles in three-dimensional space to a (picture) plane already leads to
elliptic shapes. Davé and Bashwan [27] therefore developed an algorithm
for the recognition of ellipses.

The contour of an ellipse is given by its centre point v and a positive definite, symmetric matrix A as the solution of the equation

$$(x - v)^\top A(x - v) = 1.$$

Consequently, we choose as a distance function

$$d^2(x, (v, A)) = \left(\sqrt{(x - v)^\top A(x - v)} - 1 \right)^2.$$

The matrix A contains both the lengths of the axes and the orientation of the ellipse.

Theorem 4.3 (Prototypes of AFCS) *Let* $p \in \mathbb{N}$, $D := \mathbb{R}^p$, $X = \{x_1, x_2, \ldots, x_n\} \subseteq D$, $C := \mathbb{R}^p \times \{A \in \mathbb{R}^{p \times p} \mid A$ *positive definite*$\}$, $c \in \mathbb{N}$, $R := \mathcal{P}_c(C)$, J *corresponding to* (1.7) *with* $m \in \mathbb{R}_{>1}$ *and*

$$d^2 : D \times C \to \mathbb{R}, \ (x, (v, A)) \mapsto \left(\sqrt{(x - v)^\top A(x - v)} - 1 \right)^2.$$

If J *is minimized with respect to all probabilistic cluster partitions* $X \to F(K)$ *with* $K = \{k_1, k_2, \ldots, k_c\} \in R$ *and given memberships* $f(x_j)(k_i) = u_{i,j}$ *with* $k_i = (v_i, A_i)$ *by* $f : X \to F(K)$, *then*

$$0 = \sum_{j=1}^{n} u_{i,j}^m d_{i,j}(x_j - v_i) \tag{4.4}$$

$$0 = \sum_{j=1}^{n} u_{i,j}^m d_{i,j}(x_j - v_i)(x_j - v_i)^\top \tag{4.5}$$

and

$$d_{i,j} := \frac{\sqrt{(x_j - v_i)^\top A_i(x_j - v_i)} - 1}{\sqrt{(x_j - v_i)^\top A_i(x_j - v_i)}}$$

holds.

Proof: The probabilistic cluster partition $f : X \to F(K)$ shall minimize the objective function J. Thus, the directional derivatives have to be zero at the minimum. For $i \in \mathbb{N}_{\leq c}$ and $k_i = (v_i, A_i)$ we obtain for all $\xi \in \mathbb{R}^p$:

$$0 = \frac{\partial}{\partial v_i} J$$

$$= \sum_{j=1}^{n} u_{i,j}^m \frac{\partial}{\partial v_i} \left(\sqrt{(x_j - v_i)^\top A_i(x_j - v_i)} - 1 \right)^2$$

$$= \sum_{j=1}^{n} u_{i,j}^{m} \frac{\sqrt{(x_j - v_i)^\top A_i(x_j - v_i)} - 1}{\sqrt{(x_j - v_i)^\top A_i(x_j - v_i)}} \frac{\partial}{\partial v_i}(x_j - v_i)^\top A_i(x_j - v_i)$$

$$= -\sum_{j=1}^{n} u_{i,j}^{m} d_{i,j} \left((x_j - v_i)^\top A_i \xi + \xi^\top A_i(x_j - v_i)\right)$$

$$= -2\sum_{j=1}^{n} u_{i,j}^{m} d_{i,j} \xi^\top A_i(x_j - v_i)$$

$$= -2\xi^\top \left(\sum_{j=1}^{n} u_{i,j}^{m} d_{i,j} A_i(x_j - v_i)\right).$$

The last expression must vanish for all $\xi \in \mathbb{R}^p$, therefore $A_i(\sum_{j=1}^{n} u_{i,j}^{m} d_{i,j}(x_j - v_i)) = 0$. Since A_i is positive definite and hence regular, $A_i x = 0 \Leftrightarrow x = 0$, i.e. (4.4) follows.

For the derivative with respect to the matrix A_i, the restriction to positive definite matrices causes problems. Therefore, we assume that a minimum of the objective function in the space of the positive definite matrices is also a minimum in the space of all matrices. (This condition is not at all necessarily valid. However, considering the expression $d_{i,j}$, we see that each matrix has at least to *behave like a positive definite one* for the concrete given data, in order to be able to provide real values for the square root.) Then, we can take all directional derivatives, and we obtain for all $\Delta \in \mathbb{R}^{p \times p}$:

$$0 = \frac{\partial}{\partial A_i} J$$

$$= \sum_{j=1}^{n} u_{i,j}^{m} \frac{\partial}{\partial A_i} \left(\sqrt{(x_j - v_i)^\top A_i(x_j - v_i)} - 1\right)^2$$

$$= \sum_{j=1}^{n} u_{i,j}^{m} d_{i,j} \frac{\partial}{\partial A_i}(x_j - v_i)^\top A_i(x_j - v_i)$$

$$= \sum_{j=1}^{n} u_{i,j}^{m} d_{i,j}(x_j - v_i)^\top \Delta (x_j - v_i).$$

Thus, $\nabla J(A) = \sum_{j=1}^{n} u_{i,j}^{m} d_{i,j}(x_j - v_i)(x_j - v_i)^\top = 0$, i.e. equation (4.5) is also valid. ∎

As it was already the case with the fuzzy c-shells algorithm, theorem 4.3 provides no explicit formulae for the prototypes. In their paper, Davé and Bashwan do not directly mention the algorithms with which the solutions of equations (4.4) and (4.5) should be computed. They only refer to a software package for the minimization, which also includes much more complex optimization procedures than the Newton algorithm. The solution of these equations causes a bigger problem for this algorithm than for the fuzzy c-shells algorithm, because the prototypes consist of a vector and a positive definite matrix, i.e. more parameters than the fuzzy c-shells prototypes. The application of a conventional Newton algorithm soon yields non-positive definite matrices, so that already after two iteration steps we are no longer dealing with ellipitical shapes. A naive implementation, where all matrix elements are individually optimized, does not even necessarily lead to symmetric matrices in each iteration step (and thus especially not to positive definite matrices). The search for a symmetric matrix can be quite simply achieved by optimizing only the upper or lower triangle matrix and choosing the other elements by symmetry considerations. Non-negative diagonal elements can be obtained quite easily, too, when the elements of the triangle matrix are always used as squares. This way, the search space for the Newton algorithm is restricted, but it is still a proper superset of the space of positive definite matrices.

In order to be able to still obtain results with the Newton algorithm, another representation of the matrices was chosen in our implementation. A matrix is positive definite if and only if all eigenvalues are positive. For the two-dimensional case of image recognition, the unit eigenvectors of a matrix $A \in \mathbb{R}^{2 \times 2}$ can be represented by a single real value. The two eigenvectors are perpendicular, therefore it is sufficient to consider just one parameter, the angle $\varphi \in \mathbb{R}$, determining the normalized eigenvectors in the form $e_1 = (\cos(\varphi), \sin(\varphi))^\top \in \mathbb{R}^2$ and $e_2 = (-\sin(\varphi), \cos(\varphi))^\top \in \mathbb{R}^2$. For $\lambda_1, \lambda_2 \in \mathbb{R} \setminus \{0\}$, let $\frac{1}{\lambda_1^2}$ and $\frac{1}{\lambda_2^2}$ be the associated positive eigenvalues of the normalized eigenvectors. Thus we can describe any (two-dimensional) positive definite symmetric matrix A on the basis of the parameters $(\varphi, \lambda_1, \lambda_2) \in \mathbb{R} \times (\mathbb{R} \setminus \{0\})^2$ in the form

$$A = \frac{1}{\lambda_1^2} e_1 e_1^\top + \frac{1}{\lambda_2^2} e_2 e_2^\top.$$

Since the Newton algorithm allows arbitrary real values for each parameter, contradiciting the requirement of positive eigenvalues, we use the squares of the λ-values. Because eigenvalues of a positive definite matrix are always non-zero, we also took the reciprocal of the λ-values. In this way, we ensure that the Newton algorithm must yield positive definite matrices.

However, even in this modified form the Newton algorithm is not ideal for this problem. If we try to recognize circles now, an alteration of the parameter φ causes no change in the equations, because a circle is invariant with respect to rotations. But this results in zero-column in the Jacobian matrix in the Newton algorithm. Thus, the Jacobian matrix can not be inverted, and the procedure terminates. With other iteration procedures, for instance the Levenberg-Marquardt algorithm [74, 95], results which are clearly better can be achieved here, because singular Jacobian matrices do not lead to a termination. On the other hand, these procedures have a much higher computational complexity than the comparatively primitive Newton algorithm.

Figure 4.7: AFCS analysis

The clustering results here were computed in the way described above using the Newton algorithm. For an application in practice, the usage of a more efficient procedure for the numerical solution of equation systems is recommended in order to further increase the performance of the algorithm.

Figure 4.8: AFCS analysis   Figure 4.9: AFCS analysis

When there are no ideal circles in the data set, the adaptive fuzzy c-shells algorithm can yield good results as shown in figure 4.7, when it is started with a good initialization. Also the data set in figure 4.8 with incomplete contours of ellipses was well recognized. For both figures, the fuzzy c-means algorithm was applied to obtain the initialization, already producing a rather good partition almost separating the ellipses. In addition, the radii – slightly changed to avoid circles – were estimated by equation (4.3) and used as the length for the axes of the ellipses. A good initialization is very important because otherwise the Newton algorithm does not even converge in the first step of the clustering.

The increasing number of local minima with an increasing number of clusters, and the fact that contours as cluster shapes also increase the number of local minima, affect the adaptive fuzzy c-shells algorithm, too. Figure 4.9 shows a result where the algorithm converged in a local minimum. Since the elliptic clusters are more flexible than circles, it can easily happen that a cluster covers contour segments of different ellipses in the data set. To make it even worse, other clusters are also affected by such a 'deceptive' cluster. A good initialization is consequently very important to obtain a partition corresponding to the intuition. However, in figure 4.9 the clustering result remains the same independent of an FCM or a combined FCM/FCS initialization. Even the circle clusters partition the data in almost the same way as it is done by AFCS. (Note that the ellipse on the above right is not a circle, otherwise a better result could have been expected from the circle recognition.) If we use the fuzzy c-means for the initialization, the radii of the ellipses are estimated, but they approximate only the near data in this way. This leads in our case to the partition corresponding to a local minimum. It appears to be very difficult to satisfy the demand for a good initialization with the available algorithms. For long-stretched ellipses, better results can be expected from the algorithms by Gustafson-Kessel or Gath and Geva than they could be obtained with the fuzzy c-means or the fuzzy c-(spherical) shells algorithm. For this purpose, not only the position vectors but also the orientations and radii of the recognized clusters should be determined via the eigenvalues and eigenvectors of the norm matrices. A much better approximation of the final clustering result can be expected from such an initialization.

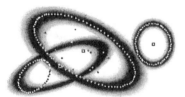
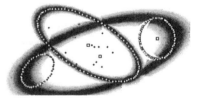

Figure 4.10: AFCS analysis   Figure 4.11: P-AFCS analysis

If the ellipses are better separated, it is easier for the adaptive fuzzy c-shells algorithm to find the correct partition. The ellipses from figure 4.10 are the same as in figure 4.9, just shifted apart a little. The AFCS algorithm recognizes – without the noise data from figure 4.10 – this data set correctly (not shown). However, the algorithm reacts very sensitively

to noise. The large ellipse that has the least Euclidean distance to the
noise data is hardly influenced. The cluster that approximates the smallest
ellipse tries to cover parts of the noise data instead. This behaviour results
from the non-Euclidean distance function of AFCS. For the same Euclidean
distance to the data, the distance function produces lower values for larger
ellipses than for smaller ones. In addition, the distance in the direction of
the axes of the ellipses is shrunk or stretched differently depending on the
shape of the ellipse. Furthermore, the data inside the ellipse always have a
distance less than one. All these factors lead to the fact that the smallest
cluster, whose main axis of the ellipse points towards the noise data, tries
to cover these additional data. With a larger number of noise data this
tendency leads to an undesirable alignment of almost all clusters so that
the partition becomes useless. A possibilistic run is unable to improve the
result. In our last example, it transforms the clustering result from figure
4.10 into the partition in figure 4.11. Because of the long axis of the *problematic cluster* from figure 4.10, the distance to the almost circular shaped
cluster (above right) is further shrunk in comparison to the Euclidean distance. Compared to the noise data, the data of the circular shaped cluster
have a high weight because of their large number. The original data of
the problematic cluster, on the other hand, get a rather low weight only,
because of their position inside the cluster. These *forces* cause the problematic cluster not to withdraw to the intuitive contour as desired, but on
the contrary to expand further. The minimization of the objective function
has a completely different meaning in the AFCS distance function than we
would intuitively expect with the Euclidean distances.

4.4   The fuzzy c-ellipsoidal shells algorithm

Another technique for the detection of ellipse-shaped clusters was developed by Frigui and Krishnapuram [35]. Their aim was to get closer to
the Euclidean distance to the contour of an ellipse by a modification of
the distance function, in order to obtain clustering results corresponding
to intuition. (For a discussion of applying FCES to the three-dimensional
case, cf. [36].) In figure 4.12, the contour of the cluster is indicated by a
circle whose centre is marked as v. The Euclidean distance of a datum x to
the cluster contour is the length of the shortest path between them. The
desired distance of the datum x from the contour of the circle corresponds
to the distance between z and x, if z is defined by the intersection point
of the cluster contour and the straight line through x and v. If we use an
ellipse instead of the circle, $\|z - x\|$ does not correspond exactly to the
Euclidean distance any more, because the straight line between x and v in-

tersects the cluster contour and is not necessarily perpendicular, and thus the connection of the contour and the datum is not the shortest path any longer. However, the expression $||z - x||$ represents a good approximation.

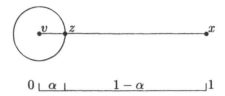

Figure 4.12: Distance measure of the FCES algorithm

Since all three points are on a straight line, there is an $\alpha \in [0, 1]$ such that $(z-v) = \alpha(x-v)$. Since z is on the ellipse, we have $(z-v)^\top A(z-v) = 1$, and thus $\alpha^2 (x - v)^\top A(x - v) = 1$. Furthermore, we obtain $x - z = x - z + v - v = (x - v) - (z - v) = (x - v) - \alpha(x - v) = (1 - \alpha)(x - v)$ or $||x - z||^2 = (1 - \alpha)^2 ||x - v||^2$. Thus, choosing $\alpha = \sqrt{\frac{1}{(x-v)^\top A(x-v)}}$ we finally derive

$$||x - z||^2 = \left(1 - \frac{1}{\sqrt{(x - v)^\top A(x - v)}} \right)^2 ||x - v||^2$$

$$= \frac{\left(\sqrt{(x - v)^\top A(x - v)} - 1 \right)^2}{(x - v)^\top A(x - v)} ||x - v||^2.$$

This expression defines the distance function for the fuzzy c-ellipsoidal shells algorithm.

Theorem 4.4 (Prototypes of FCES) *Let $p \in \mathbb{N}$, $D := \mathbb{R}^p$, $X = \{x_1, x_2, \ldots, x_n\} \subseteq D$, $C := \mathbb{R}^p \times \{A \in \mathbb{R}^{p \times p} \mid A \text{ positive definite}\}$, $c \in \mathbb{N}$, $R := \mathcal{P}_c(C)$, J corresponding to (1.7) with $m \in \mathbb{R}_{>1}$ and*

$$d^2 : D \times C \to \mathbb{R}, \ (x, (v, A)) \mapsto \frac{\left(\sqrt{(x - v)^\top A(x - v)} - 1 \right)^2 ||x - v||^2}{(x - v)^\top A(x - v)}.$$

If J is minimized with respect to all probabilistic cluster partitions $X \to F(K)$ with $K = \{k_1, k_2, \ldots, k_c\} \in R$ and given memberships $f(x_j)(k_i) = u_{i,j}$ with $k_i = (v_i, A_i)$ by $f : X \to F(K)$, then

$$0 \ = \ \sum_{j=1}^{n} \frac{u_{i,j}^m \left(\sqrt{d_{i,j}} - 1\right)}{d_{i,j}^2}$$

$$\cdot \left[||x_j - v_i||^2 A_i + \left(\sqrt{d_{i,j}} - 1\right) d_{i,j} I\right] (x_j - v_i) \qquad (4.6)$$

$$0 \ = \ \sum_{j=1}^{n} u_{i,j}^m (x_j - v_i)(x_j - v_i)^\top \left(\frac{||x_j - v_i||}{d_{i,j}}\right)^2 \left(\sqrt{d_{i,j}} - 1\right), \qquad (4.7)$$

holds where I *denotes the identity matrix of* $\mathbb{R}^{p \times p}$ *and*

$$d_{i,j} := (x_j - v_i)^\top A_i (x_j - v_i)$$

for all $j \in \mathbb{N}_{\leq n}$ *and* $i \in \mathbb{N}_{\leq c}$.

Proof: The probabilistic cluster partition $f : X \to F(K)$ shall minimize the objective function J. Let $f_{i,j} := \left(\sqrt{d_{i,j}} - 1\right)^2 ||x_j - v_i||^2$ and $g_{i,j} := \frac{1}{d_{i,j}}$

for all $j \in \mathbb{N}_{\leq n}$ and $i \in \mathbb{N}_{\leq c}$. Thus $J = \sum_{j=1}^{n} \sum_{i=1}^{c} u_{i,j}^m f_{i,j} g_{i,j}$. The minimality of J implies that the directional derivatives must vanish. Hence, we have for all $i \in \mathbb{N}_{\leq c}$ and $\xi \in \mathbb{R}^p$:

$$\frac{\partial f_{i,j}}{\partial v_i} = \frac{\partial}{\partial v_i} \left(\sqrt{d_{i,j}} - 1\right)^2 (x_j - v_i)^\top I (x_j - v_i)$$

$$= -\frac{2\left(\sqrt{d_{i,j}} - 1\right)}{2\sqrt{d_{i,j}}} 2\xi^\top A_i (x_j - v_i)(x_j - v_i)^\top I (x_j - v_i)$$

$$- \left(\sqrt{d_{i,j}} - 1\right)^2 2\xi^\top I (x_j - v_i)$$

$$= -2\xi^\top \left[A_i (x_j - v_i) ||x_j - v_i||^2 \frac{\sqrt{d_{i,j}} - 1}{\sqrt{d_{i,j}}} + \right.$$

$$\left. I (x_j - v_i) \left(\sqrt{d_{i,j}} - 1\right)^2 \right].$$

$$\frac{\partial g_{i,j}}{\partial v_i} = \frac{\partial}{\partial v_i} \frac{1}{d_{i,j}}$$

$$= \frac{1}{d_{i,j}^2} 2\xi^\top A_i (x_j - v_i).$$

$$0 = \frac{\partial J}{\partial v_i} = \sum_{j=1}^{n} u_{i,j}^m \left(\frac{\partial f_{i,j}}{\partial v_i} g_{i,j} + \frac{\partial g_{i,j}}{\partial v_i} f_{i,j}\right)$$

$$= 2\xi^\top \sum_{j=1}^{n} u_{i,j}^m \left[A_i(x_j - v_i) \|x_j - v_i\|^2 \cdot \right.$$

$$\left. \left(-\frac{\sqrt{d_{i,j}} - 1}{d_{i,j}\sqrt{d_{i,j}}} + \frac{(\sqrt{d_{i,j}} - 1)^2}{d_{i,j}^2} \right) - \frac{I(x_j - v_i)(\sqrt{d_{i,j}} - 1)^2}{d_{i,j}} \right]$$

$$= -2\xi^\top \sum_{j=1}^{n} \frac{u_{i,j}^m (\sqrt{d_{i,j}} - 1)}{d_{i,j}^2} \left[A_i(x_j - v_i) \|x_j - v_i\|^2 \right.$$

$$\left. + I(x_j - v_i) \left(\sqrt{d_{i,j}} - 1 \right) d_{i,j} \right].$$

Since the last expression must be zero for all $\xi \in \mathbb{R}^p$, we obtain

$$0 = \sum_{j=1}^{n} \frac{u_{i,j}^m (\sqrt{d_{i,j}} - 1)}{d_{i,j}^2} \left[\|x_j - v_i\|^2 A_i + \left(\sqrt{d_{i,j}} - 1 \right) d_{i,j} I \right] (x_j - v_i).$$

This corresponds to equation (4.6). If we take the directional derivatives in $\mathbb{R}^{p \times p}$, we have for all $\Delta \in \mathbb{R}^{p \times p}$:

$$\frac{\partial f_{i,j}}{\partial A_i} = \frac{\partial}{\partial A_i} \|x_j - v_i\|^2 \left(\sqrt{d_{i,j}} - 1 \right)^2$$

$$= \|x_j - v_i\|^2 \frac{\sqrt{d_{i,j}} - 1}{\sqrt{d_{i,j}}} (x_j - v_i)^\top \Delta (x_j - v_i)$$

$$\frac{\partial g_{i,j}}{\partial A_i} = \frac{\partial}{\partial A_i} \frac{1}{d_{i,j}}$$

$$= -\frac{(x_j - v_i)\Delta(x_j - v_i)}{d_{i,j}^2}.$$

$$0 = \frac{\partial b}{\partial A_i} = \sum_{j=1}^{n} u_{i,j}^m \left(\frac{\partial f_{i,j}}{\partial A_i} g_{i,j} + \frac{\partial g_{i,j}}{\partial A_i} f_{i,j} \right)$$

$$= \sum_{j=1}^{n} u_{i,j}^m \frac{(x_j - v_i)^\top \Delta (x_j - v_i)}{d_{i,j}} \left[\frac{\|x_j - v_i\|^2 (\sqrt{d_{i,j}} - 1)}{\sqrt{d_{i,j}}} \right.$$

$$\left. - \frac{\|x_j - v_i\|^2 (\sqrt{d_{i,j}} - 1)^2}{d_{i,j}} \right]$$

$$= \sum_{j=1}^{n} u_{i,j}^m (x_j - v_i)^\top \Delta (x_j - v_i) \left(\frac{\|x_j - v_i\|}{d_{i,j}} \right)^2 \left(\sqrt{d_{i,j}} - 1 \right).$$

In the same way as in the proof of theorem 4.3, we obtain equation (4.7).

∎

As was the case for the adaptive fuzzy c-shells algorithm, theorem 4.4 provides no explicit equations for the prototypes. Again, a numerical iteration scheme is needed for the solution of the system of equations. For the same reasons as in section 4.3, convergence cannot be guaranteed in all cases using a simple Newton algorithm, because the Newton iteration terminates when the Jacobian matrix is singular. Such singularities can easily occur when the minimum (or saddle point) has already been reached in a component.

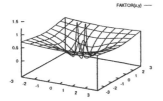
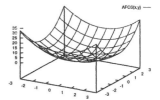

Figure 4.13: Distance function   Figure 4.14: Distance function
FCES without a Euclidean factor   AFCS

With respect to complexity, FCES gives no improvement compared to AFCS. But the the modified distance function has certain advantages. Different distances are produced by the AFCS distance function for long-stretched ellipses as well as for ellipses of a different size. The fuzzy c-ellipsoidal shells algorithm does not suffer from these drawbacks.

Figure 4.13 shows a graph of $\frac{\left(\sqrt{x^\top A x}-1\right)^2}{x^\top A x}$, i.e. the FCES distance function without the factor of the Euclidean norm. The intersection with a plane at the height 0 gives the contour of the ellipse. Inside the ellipse, the function increases towards infinity, outside it converges towards 1. There, the multiplication with the Euclidean norm produces approximately Euclidean distances. That means a great improvement compared to the distance function of the AFCS algorithm, which is shown in figure 4.14. Within the small area illustrated, the distance function already produces values above 30, clearly more than the square of the Euclidean distance from the centre point of the ellipse ($3^2 + 3^2 = 18$), and even higher values for the distance to the contour of the ellipse. Scaled like this, the ellipse itself cannot be seen any longer, because the distance function is always one in the centre of the ellipse.

Figure 4.15 shows the distance functions of AFCS and FCES, and the function of figure 4.13, from top to bottom. The differences between the

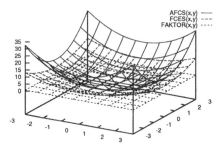

Figure 4.15: Distance function of the AFCS and FCES, as well as the FCES factor from figure 4.13

AFCS and FCES distance functions are considerable. The actual Euclidean distance from the contour of the ellipse is quite close to the FCES distance function.

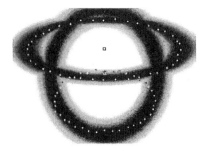

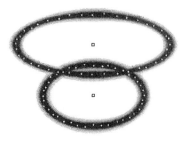

Figure 4.16: AFCS analysis   Figure 4.17: FCES analysis

Figures 4.16 and 4.17 show a data set where the changed distance function led to a better partition. The data near the intersection of the two ellipses are completely assigned to the upper ellipse by AFCS. This ellipse minimizes the distances by taking its course between the actual contours of the ellipses. However, the memberships of these data are quite high for the cluster of the lower ellipse, too, because they can have a maximum distance of 1. But the distances are not large enough to draw the cluster into the correct form. The FCES algorithm solves this problem and the ellipses are recognized correctly. That the result of FCES was not just a lucky strike, can be shown by the fact that even an FCES run initialized with the good AFCS partition leads again to figure 4.17. The modified distance function has a positive influence on the analysis result in this example. (Vice versa, the AFCS algorithm remains at the result of FCES with an initialization with the FCES algorithm. This clustering result is hence at least a local minimum for the AFCS distance function, too.)

Figure 4.18: FCES analysis   Figure 4.19: P-FCES analysis

However, it is not true that FCES always meets the intuitive partition better. The data set in figure 4.18 was correctly partitioned by AFCS, as shown in figure 4.8. In the FCES run, the cluster in the middle obviously converged *faster* and covered some data from the cluster on the right. The algorithm converges into a local minimum. If probabilistic FCES is released from the competition among the clusters and carries out a possibilistic clustering, the wrongly partitioned data are assigned to the cluster on the right. The analysis result from figure 4.19 thus corresponds better to intuition. (The correction by possibilistic clustering does not work with AFCS and the data from figure 4.16, though. Possiblistic clustering does not principally achieve a partition which better corresponds to intuition.)

For long-stretched ellipses, the membership degrees are not uniform along the contour of the ellipse (cf. 4.18, for instance). This effect is caused by the distance function. This shows again that we do not have an exact Euclidean distance yet. This fact may seem a little puzzling in the graphical representation, but it does not influence the results of the analysis noticeably.

There is also a drawback in the distance function of the fuzzy c-ellipsoidal shells algorithm, though. In equation (4.7), there is a division by $d_{i,j}^2$. With positive definite matrices A with very large elements, the $d_{i,j}^2$ grow strongly. When searching for roots of (4.7), the Newton algorithm thus sometimes tends to matrices that become larger and larger, leading to an imaginary solution in infinity. As a matter of fact, the cluster degenerates here to a very small ellipse that has no meaning for image recognition.

4.5 The fuzzy c-ellipses algorithm

The algorithms for the detection of ellipses discussed so far require numerical solutions of the equations for the prototypes. With the fuzzy c-ellipses (FCE) [38], Gath and Hoory introduced an algorithm for the recognition of ellipses that uses a nearly Euclidean distance function and contains explicit equations for the prototypes. However, the prototypes do not minimize the objective function with respect to the given memberships. If at all, a small value is obtained for the objective function, but not necessarily a (local) minimum. The prerequisites for the convergence of the fuzzy clustering techniques, which were formulated in a relatively general manner by Bezdek, are thus not fulfilled. The FCE algorithm deviates from other algorithms distinctly in that point. However, the high computational complexity of AFCS and FCES motivate the consideration of the FCE algorithm. Moreover, we already mentioned that for the Newton algorithm also, it was not necessary to compute the exact solution (minimum) in order to obtain good results [12]. (This also confirms the subjective impression that the FCE algorithm needs more iteration steps than a comparable AFCS or FCES run.)

An ellipse prototype of FCE is characterized by two focal points v^0, v^1, and a radius r. The distance between a datum x and an ellipse is defined by $| \, ||x - v^0|| + ||x - v^1|| - r \, |$. The distance function defined this way is not only close to the Euclidean distance measure, but it is also rather simply structured compared to the FCES distance. Instead of the centre form used before, the focal point form is used here, as it is applied in Kepler's Laws, for instance.

Remark 4.5 (Prototypes of FCE)
Let $D := \mathbb{R}^2$, $X = \{x_1, x_2, \ldots, x_n\} \subseteq D$, $C := \mathbb{R}^2 \times \mathbb{R}^2 \times \mathbb{R}$, $c \in \mathbb{N}$, $R := \mathcal{P}_c(C)$, J corresponding to (1.7) with $m \in \mathbb{R}_{>1}$ and

$$d^2 : D \times C \to \mathbb{R}, \quad (x, (v^0, v^1, r)) \mapsto (\, ||x - v^0|| + ||x - v^1|| - r \,)^2.$$

If J is minimized by $f : X \to F(K)$ with respect to all probabilistic cluster partitions $X \to F(K)$ with $K = \{k_1, k_2, \ldots, k_c\} \in E$ and given memberships $f(x_j)(k_i) = u_{i,j}$ with $k_i = (v_i^0, v_i^1, r_i)$ and $l \in \{0, 1\}$, then

$$0 \;=\; -2 \sum_{j=1}^{n} u_{i,j}^m (||x_j - v_i^0|| + ||x_j - v_i^1|| - r_i), \tag{4.8}$$

$$0 \;=\; -2 \sum_{j=1}^{n} u_{i,j}^m \left[(x_j - v_i^l) + (||x_j - v_i^{1-l}|| - r_i) \frac{x_j - v_i^l}{||x_j - v_i^l||} \right] \tag{4.9}$$

holds.

Equations (4.8) and (4.9) are directly obtained by the partial derivatives with respect to the radius and the vector components. Since equations (4.8) and (4.9) are not explicit, they require a numerical solution. However, in the FCE algorithm the updated radius r_i is computed by transforming (4.8) with respect to r_i and inserting the old focal points:

$$r_i = \frac{\sum_{j=1}^{n} u_{i,j}^m (\|x_j - v_i^0\| + \|x_j - v_i^1\|)}{\sum_{j=1}^{n} u_{i,j}^m}.$$

Then equation (4.9) is transformed into

$$v_i^l = \frac{\sum_{j=1}^{n} u_{i,j}^m \left(x_j + (\|x_j - v_i^{1-l}\| - r_i)\frac{x_j - v_i^l}{\|x_j - v_i^l\|}\right)}{\sum_{j=1}^{n} u_{i,j}^m}$$

and evaluated with the just computed r_i and the old focal points. This heuristic method can of course also be applied to other shell clustering algorithms like FCS or AFCS, for which no explicit equations for the prototypes exist. And indeed, as already mentioned, the fuzzy c-rings algorithm by Man and Gath [72] uses the equations in the same manner. However, convergence has not been proved for this technique.

To compute the new prototypes, we have to initialize the FCE prototypes, because they are already needed in the first iteration step. Initializing just the memberships, for which we could use FCM, is not sufficient. Gath and Hoory therefore use the following initialization: After ten FCM steps, the fuzzy covariance matrix

$$S_i = \frac{\sum_{j=1}^{n} u_{i,j}^m (x_j - v_i)(x_j - v_i)^\top}{\sum_{j=1}^{n} u_{i,j}^m}$$

is calculated where v_i is the FCM centre of cluster i. The eigenvectors e_i^0 and e_i^1 and the corresponding eigenvalues λ_i^0 and λ_i^1 are derived from S_i. Each eigenvalue corresponds to the square of the length of an ellipse axis. Let us assign the eigenvalue and eigenvector of the main axis the index 0. (The main axis has the largest eigenvalue.) Then, the focal points are $v_i^0 = v_i + f_i \cdot e_i^0$ and $v_i^1 = v_i - f_i \cdot e_i^0$ where $f_i = \frac{\sqrt{\lambda_i^0 - \lambda_i^1}}{2}$. The radius r_i is initialized with $\sqrt{\lambda_i^0}$, the length of the ellipse's main axis. For the data set from figure 4.7, the initialization procedure is already sufficient for the recognition of the five ellipses. For separated ellipses, the procedure of the FCE algorithm seems to be sufficient in each case. (If it is known that the ellipse clusters are interlaced, Gath and Hoory recommend to scatter c circle clusters with different radii around the centre of gravity of all data and use this as an initialization instead of FCM.)

Figure 4.20: FCE analysis   Figure 4.21: P-FCE analysis

But even for examples that could not be clustered correctly by the AFCS and the FCES algorithms, the FCE algorithm does not perform any worse as is shown in figures 4.20 and 4.21. The ellipses' segments do not meet the original ellipses exactly but are better than with the two other algorithms. The data set from 4.9 is clustered by FCE in the same (wrong) way as it was for the case with AFCS before (figure 4.21). The similarity of the results is nevertheless remarkable considering the enormously simplified computation scheme. In addition, the distortion caused by noise seems to be reduced for FCE compared to AFCS. Because of the distance, the outlier cluster is fixed by two data on the large contour and not distorted by the noise data farther to the right.

4.6 The fuzzy c-quadric shells algorithm

The disadvantage of the shell clustering algorithms so far for the recognition of elliptic contours is caused by the *numerical* solution of a non-linear system of equations, or the heuristic method for computing the prototypes. The fuzzy c-quadric shells algorithm does not need such procedures, since explicit equations for the prototypes can be provided. Moreover, the fuzzy c-quadric shells algorithm [64, 65, 66] is in principle able to recognize – besides contours of circles and ellipses – hyperbolas, parabolas or linear clusters. The prototypes are second degree polynomials whose roots (cluster shapes) are quadrics. In the two-dimensional case of image recognition, the distance of a datum (x_1, x_2) from the prototype $a = (a_1, a_2, \ldots, a_6) \in \mathbb{R}^6$ follows directly from the quadratic polynomial

$$a_1 x_1^2 + a_2 x_2^2 + a_3 x_1 x_2 + a_4 x_1 + a_5 x_2 + a_6$$

which is a measure for the distance to the root and thus to the cluster contour. (The quadratic polynomial is written in matrix form in theorem 4.6.) In the case of $\mathbb{R}^2$, the quadrics are just the conic sections.

The minimization of the distance function has the undesired trivial solution $a = 0$. This can be prevented by additional constraints. Krishnapuram, Frigui and Nasraoui [64, 65, 66] suggest $\|(a_1, a_2, a_3)\|^2 = 1$. Although this constraint excludes linear clusters in the two-dimensional case, it is often used because it yields closed prototype update equations. The algorithm approximates straight lines by hyperbolas or parabolas.

Theorem 4.6 (Prototypes of FCQS) *Let* $p \in \mathbb{N}$, $D := \mathbb{R}^p$, $X = \{x_1, x_2, \ldots, x_n\} \subseteq D$, $r := \frac{p(p+1)}{2}$, $C := \mathbb{R}^{r+p+1}$, $c \in \mathbb{N}$, $R := \mathcal{P}_c(C)$, J corresponding to (1.7) with $m \in \mathbb{R}_{>1}$,*

$$\Phi : C \to \mathbb{R}^{p \times p},$$

$$(a_1, a_2, \ldots, a_{r+p+1}) \mapsto \begin{pmatrix} a_1 & \frac{1}{2}a_{p+1} & \frac{1}{2}a_{p+2} & \cdot & \cdot & \frac{1}{2}a_{2p-1} \\ \frac{1}{2}a_{p+1} & a_2 & \frac{1}{2}a_{2p} & \cdot & \cdot & \frac{1}{2}a_{3p-3} \\ \frac{1}{2}a_{p+2} & \frac{1}{2}a_{2p} & a_3 & \cdot & \cdot & \frac{1}{2}a_{4p-6} \\ \cdot & \cdot & \cdot & \cdot & \cdot & \cdot \\ \cdot & \cdot & \cdot & \cdot & \cdot & \frac{1}{2}a_r \\ \frac{1}{2}a_{2p-1} & \frac{1}{2}a_{3p-3} & \frac{1}{2}a_{4p-6} & \cdot & \frac{1}{2}a_r & a_p \end{pmatrix}$$

and

$$d^2 : D \times C \to \mathbb{R},$$
$$(x, a) \mapsto \left(x^\top \Phi(a) x + x^\top (a_{r+1}, a_{r+2}, \ldots, a_{r+p})^\top + a_{r+p+1}\right)^2.$$

If J is minimized under the constraint $\|p_i\| = 1$ by $f : X \to F(K)$ with $p_i = (k_{i,1}, k_{i,2}, \ldots, k_{i,p}, \frac{k_{i,p+1}}{\sqrt{2}}, \frac{k_{i,p+2}}{\sqrt{2}}, \ldots, \frac{k_{i,r}}{\sqrt{2}})$ and $q_i = (k_{i,r+1}, k_{i,r+2}, \ldots, k_{i,p+1})$ with respect to all probabilistic cluster partitions $X \to F(K)$ with $K = \{k_1, k_2, \ldots, k_c\} \in E$ and given memberships $f(x_j)(k_i) = u_{i,j}$, then

$$p_i \quad \text{is eigenvector of } R_i - T_i^\top S_i^{-1} T_i \text{ with} \tag{4.10}$$
$$\text{a minimal eigenvalue}$$

$$q_i \quad = \quad -S_i^{-1} T_i p_i, \tag{4.11}$$

where

$$R_i = \sum_{j=1}^n u_{i,j}^m \tilde{R}_j, \quad S_i = \sum_{j=1}^n u_{i,j}^m \tilde{S}_j, \quad T_i = \sum_{j=1}^n u_{i,j}^m \tilde{T}_j$$

and for all $j \in \mathbb{N}_{\leq n}$

$$\tilde{R}_j \;=\; r_j r_j^\top, \quad \tilde{S}_j = s_j s_j^\top, \quad \tilde{T}_j = s_j r_j^\top$$

$$r_j^\top \;=\; (x_{j,1}^2, x_{j,2}^2, ..., x_{j,p}^2, \sqrt{2}x_{j,1}x_{j,2}, \sqrt{2}x_{j,1}x_{j,3}, ..., \sqrt{2}x_{j,1}x_{j,p},$$
$$\sqrt{2}x_{j,2}x_{j,3}, \sqrt{2}x_{j,2}x_{j,4}, ..., \sqrt{2}x_{j,p-1}x_{j,p})$$

$$s_j^\top \;=\; (x_{j,1}, x_{j,2}, ...x_{j,p}, 1).$$

Proof: The probabilistic cluster partition $f : X \to F(K)$ shall minimize the objective function J under the constraint $||p_i|| = 1$. Similar to the fuzzy c-spherical shells algorithm, the distance function can be rewritten in the form $d^2(x_j, k_i) = k_i^\top \tilde{M}_j k_i$, where $\tilde{M}_j := \begin{pmatrix} \tilde{R}_j & \tilde{T}_j^\top \\ \tilde{T}_j & \tilde{S}_j \end{pmatrix}$ for all $j \in \mathbb{N}_{\leq n}$. If the constraint is taken into account by Lagrange multipliers, the objective function thus becomes $J = \sum_{j=1}^n \sum_{i=1}^c u_{i,j}^m \, k_i^\top \tilde{M}_j k_i - \lambda_i (||p_i|| - 1)$. The minimality of J implies that the directional derivatives of J with respect to k must be zero. Therefore, we have for all $\xi \in \mathbb{R}^{r+p+1}$ with $M_i := \sum_{j=1}^n u_{i,j}^m \tilde{M}_j$

$$
\begin{aligned}
0 &= \frac{\partial J}{\partial k_i} \\
&= \frac{\partial}{\partial k_i} \left(k_i^\top M_i k_i - \lambda_i (||p_i|| - 1) \right) \\
&= -2\xi^\top M_i k_i - 2\lambda_i (\xi_1, \xi_2, \ldots, \xi_r)^\top p_i \\
&= -2\xi^\top \left(M_i k_i - \lambda_i \begin{pmatrix} p_i \\ 0 \end{pmatrix} \right).
\end{aligned}
$$

Since the last expression must be zero for all $\xi \in \mathbb{R}^{r+p+1}$, we obtain

$$M_i k_i \;=\; \lambda_i \begin{pmatrix} p_i \\ 0 \end{pmatrix} \qquad\qquad (4.12)$$

$$\Leftrightarrow \quad \begin{pmatrix} R_i & T_i^\top \\ T_i & S_i \end{pmatrix} \begin{pmatrix} p_i \\ q_i \end{pmatrix} \;=\; \lambda_i \begin{pmatrix} p_i \\ 0 \end{pmatrix}$$

$$\Leftrightarrow \quad \begin{pmatrix} R_i p_i + T_i^\top q_i \\ T_i p_i + S_i q_i \end{pmatrix} \;=\; \lambda_i \begin{pmatrix} p_i \\ 0 \end{pmatrix}$$

$$\Leftrightarrow \quad \begin{pmatrix} R_i p_i + T_i^\top q_i \\ q_i \end{pmatrix} \;=\; \begin{pmatrix} \lambda_i p_i \\ -S_i^{-1} T_i p_i \end{pmatrix}$$

$$\Leftrightarrow \quad \begin{pmatrix} \left(R_i - T_i^\top S_i^{-1} T_i \right) p_i \\ q_i \end{pmatrix} \;=\; \begin{pmatrix} \lambda_i p_i \\ -S_i^{-1} T_i p_i \end{pmatrix} \quad .$$

The lower part of the last equation is identical to (4.11). The upper part has r solutions, viz. the eigenvectors and the corresponding eigenvalues of the matrix $R_i - T_i^\top S_i^{-1} T_i$ for p_i and λ_i. From the minimality of J at f follows from the derivation to λ_i that the constraint $\|p_i\| = 1$ is valid. Thus with equation (4.12), we have for the objective function:

$$J = \sum_{j=1}^{n} \sum_{i=1}^{c} u_{i,j}^m \; k_i^\top \lambda_i \begin{pmatrix} p_i \\ 0 \end{pmatrix} = \lambda_i \sum_{j=1}^{n} \sum_{i=1}^{c} u_{i,j}^m \; p_i^\top p_i = \lambda_i \sum_{j=1}^{n} \sum_{i=1}^{c} u_{i,j}^m \; .$$

From the possible eigenvalue/eigenvector combinations, we choose the one with the smallest eigenvalue with the corresponding normalized eigenvector so that (4.10) is also proved. ■

Figure 4.22: FCQS analysis   Figure 4.23: FCQS analysis

Krishnapuram, Frigui and Nasraoui [66] recommend initializing the fuzzy c-quadric shells algorithm by ten fuzzy c-means steps with $m = 3$, ten Gustafson-Kessel steps with $m = 3$ and five fuzzy c-shells steps with $m = 2$. Two clustering results of FCQS are shown in figures 4.22 and 4.23. In both cases, the initialization provides a very good approximation for the final cluster partition. Correspondingly, the results of the algorithm are very good. The two intersecting lines in figure 4.22 are clustered by a single quadric (a hyperbola). If we search for three clusters, the shown clusters will be obtained again, and the new cluster will be a parabola that does not provide any additional information.

Figure 4.24 shows an ellipse, a circle, a straight line, and a parabola, that were also recognized well by the fuzzy c-quadric shells algorithm. If the goal was a just correct assignment of the data to clusters, the result would be very good. A similar result is shown in figure 4.25. In both figures, the shapes of straight lines and parabolas are represented by hyperbolas so that a post-processing of the clustering results is necessary. Hyperbolas are quite unusual patterns in image recognition so a plausibility check can be performed automatically when shapes like these occur. Krishnapuram [66] provides table A.3 (cf. appendix) which allows us to estimate cluster

Figure 4.24: FCQS analysis   Figure 4.25: FCQS analysis

shapes from the FCQS result. He assumes that hyperbolas, parabolas or extremely long-stretched ellipses correspond to segments of straight lines in the picture. Since the fuzzy c-quadric shells algorithm can not recognize straight lines directly by definition, he suggests an algorithm for subsequent detections of lines.

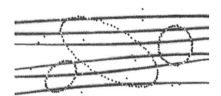

Figure 4.26: FCQS analysis   Figure 4.27: FCQS analysis

All test patterns used so far were artificially generated. Data from a *real* picture which were preprocessed by contour operators can have a somewhat different *structure*. A straight-forward procedure for generating artificial ellipses data is to insert pixels near the contour at equidistant angles. This means that the pixel density along the main axis is higher than along the shorter axis. This irregularity can also be observed in some other work [27]. In some papers about the fuzzy c-quadric shells algorithm, the point density is higher at the shorter axis. Figure 4.26 shows a data set like that. Figure 4.27 shows the same ellipses with additional data near the short axes of the ellipses. This small change drastically influences the analysis result. In figure 4.26, many data accumulate on a relatively short segment of the ellipse. These regions with a high data density can be minimized locally by the objective function. The whole data set can be covered by long-stretched ellipses or hyperbolas, where only the regions with a high data density are matched by the contours of the clusters. The result is useless for image recognition. In figure 4.27, the contours of the ellipses are clearly better recognized, they are only distorted by the surrounding noise data, especially by those *outside* the contour. The fuzzy c-quadric shells algorithm thus reacts very sensitively to the distribution of the data along the contour. This has to be considered for preprocessing the image. However, we can usually assume an equidistant distribution of the data along the ellipse contours.

In some older publications on shell clustering algorithms, the figures only show which data are grouped in a cluster. However, especially with the fuzzy c-quadric shells algorithm, it is much more crucial for image recognition whether the cluster parameters have been recognized correctly. For example, figure 4.27 shows that the correct determination of these parameters does not necessarily follow from a correct partition of the data.

Figure 4.28: FCQS analysis

Possibilistic clustering can also be applied to the fuzzy c-quadric shells algorithm, however, the desired effect is only achieved if the clusters do not deviate too much from the intuitive partition. Because of the multitude of possible cluster shapes, FCQS sometimes determines a very bizarre shape half-covering the actual cluster as well as noise data – as in figure 4.28 where the clusters' shape was mainly influenced by (only) two outliers (left on top and right on bottom). In this case, possibilistic clustering does not improve the result.

Unfortunately, the distance function of the FCQS algorithm is highly non-linear, like the AFCS distance function. For the same Euclidean distance, the FCQS distance function produces greater values for smaller quadrics than for larger ones, and inside a quadric smaller distances than outside. Since the intuitive partition results from considering the Euclidean distance, a reason for the unsatisfactory results might be the distance function.

4.7 The modified fuzzy c-quadric shells algorithm

Because of the inadequacies of the FCQS distance function, Krishnapuram et al. suggest a modification like that of the AFCS algorithm. Once again, z denotes the point that results from the intersection of the quadric and the straight line between the datum x and the centre of the quadric (cf. figure 4.12). In particular, the distance of the point z from the quadric is zero. Hence, in the two-dimensional case, we have for $k \in \mathbb{R}^6$:

$$z^\top A z + z^\top b + c = 0 \quad \text{with} \quad A = \begin{pmatrix} k_1 & \frac{1}{2}k_3 \\ \frac{1}{2}k_3 & k_2 \end{pmatrix}, \; b = \begin{pmatrix} k_4 \\ k_5 \end{pmatrix}, \; c = k_6.$$

In order to simplify the computation of z, we first rotate the quadric by an angle α such that its axes are parallel to the coordinate axes, i.e. A becomes a diagonal matrix. The distance between the points x and z (which are both rotated as well) is not changed by the rotation. Thus, let

$$R = \begin{pmatrix} \cos(\alpha) & \sin(\alpha) \\ -\sin(\alpha) & \cos(\alpha) \end{pmatrix},$$

$$\tilde{A} = \begin{pmatrix} a_1 & a_3 \\ a_3 & a_2 \end{pmatrix} = R^\top A R,$$

$$\tilde{b} = \begin{pmatrix} a_4 \\ a_5 \end{pmatrix} = R^\top b \quad \text{and}$$

$$\tilde{c} = a_6 = c.$$

From the requirement of a diagonal matrix, it follows together with the trigonometric addition theorems:

$$0 = a_3$$

$$= \cos(\alpha)\sin(\alpha)(k_1 - k_2) + \frac{k_3}{2}(\cos^2(\alpha) - \sin^2(\alpha))$$

$$\Leftrightarrow \quad (k_1 - k_2)\sin(2\alpha) = -\frac{k_3}{2}\,2\cos(2\alpha)$$

$$\Leftrightarrow \quad \frac{\sin(2\alpha)}{\cos(2\alpha)} = \frac{-k_3}{k_1 - k_2}$$

$$\Leftrightarrow \quad \alpha = \frac{1}{2}\text{atan}\left(\frac{-k_3}{k_1 - k_2}\right).$$

We denote the likewise rotated points z and x by $\tilde{z} = R^\top z$ and $\tilde{x} = R^\top x$. The distance between $\tilde{z}$ and $\tilde{x}$ can be written as the minimum distance

$\|\tilde{x} - \tilde{z}\|^2$ under the constraint $\tilde{z}^\top \tilde{A} \tilde{z} + \tilde{z}^\top \tilde{b} + \tilde{c} = 0$. Using a Lagrange multiplier λ, we obtain for all $\xi \in \mathbb{R}^2$:

$$
\begin{aligned}
0 &= \frac{\partial}{\partial \tilde{z}} \left[(\tilde{x} - \tilde{z})^\top (\tilde{x} - \tilde{z}) - \lambda (\tilde{z}^\top \tilde{A} \tilde{z} + \tilde{z}^\top \tilde{b} + \tilde{c}) \right] \\
&= -2\xi^\top (\tilde{x} - \tilde{z}) - 2\lambda \xi^\top \tilde{A} \tilde{z} - \lambda \xi^\top \tilde{b} \\
&= \xi^\top \left[-2\tilde{x} - \lambda \tilde{b} + 2(\tilde{z} - \lambda \tilde{A} \tilde{z}) \right] .
\end{aligned}
$$

Since the last expression must be zero for all $\xi \in \mathbb{R}^2$, it further follows that

$$
\begin{aligned}
2\tilde{x} + \lambda \tilde{b} &= 2(I - \lambda \tilde{A}) \tilde{z} \\
\Leftrightarrow \quad \tilde{z} &= \frac{1}{2} (I - \lambda \tilde{A})^{-1} (2\tilde{x} + \lambda \tilde{b}) .
\end{aligned}
$$

Since $\tilde{A}$ is a diagonal matrix, $(I + \lambda \tilde{A})$ can be easily inverted so that we have the following result for $\tilde{z}$:

$$
\tilde{z}^\top = \left(\frac{\lambda a_4 + 2\tilde{x}_1}{2(1 - \lambda a_1)}, \frac{\lambda a_5 + 2\tilde{x}_2}{2(1 - \lambda a_2)} \right) .
$$

If we insert this expression for $\tilde{z}$ into the constraint, we obtain a fourth degree polynomial in the Lagrange multiplier λ whose roots can be computed in a closed form:

$$
C_4 \lambda^4 + C_3 \lambda^3 + C_2 \lambda^2 + C_1 \lambda + C_0 = 0.
$$

The coefficients are:

$$
\begin{aligned}
C_4 &= a_1 a_2 (4 a_1 a_2 a_6 - a_2 a_4^2 - a_1 a_5^2), \\
C_3 &= 2 a_1 a_2 (a_4^2 + a_5^2) + 2(a_1^2 a_5^2 + a_2^2 a_4^2) - 8 a_1 a_2 a_6 (a_1 + a_2), \\
C_2 &= 4 a_1 a_2 (a_2 \tilde{x}_1^2 + a_1 \tilde{x}_2^2) - a_1 a_4^2 - a_2 a_5^2 \\
&\quad + 4 a_2 a_4 (a_2 \tilde{x}_1 - a_4) + 4 a_1 a_5 (a_1 \tilde{x}_2 - a_5) + 4 a_6 (a_1^2 + a_2^2 + 4 a_1 a_2), \\
C_1 &= -8 a_1 a_2 (\tilde{x}_1^2 + \tilde{x}_2^2) + 2(a_4^2 + a_5^2) \\
&= -8 (a_2 a_4 \tilde{x}_1 + a_1 a_5 \tilde{x}_2) - 8 a_6 (a_1 + a_2), \\
C_0 &= 4 (a_1 \tilde{x}_1^2 + a_2 \tilde{x}_2^2 + a_4 \tilde{x}_1 + a_5 \tilde{x}_2 + a_6) .
\end{aligned}
$$

For each of the up to four real solutions λ_i, $i \in \mathbb{N}_{\leq 4}$, we obtain a solution for $\tilde{z}_i$. The minimization problem is then reduced to $d = \min\{ \|x - z_i\|^2 \mid i \in \mathbb{N}_{\leq 4} \}$.

The minimization of the objective function with this modified distance function requires numerical procedures. In order to avoid this additional

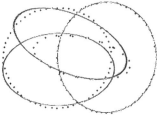

Figure 4.29: FCQS analysis   Figure 4.30: MFCQS analysis

computational burden, we compute the prototypes according to theorem 4.6 despite the changed distance function, and *hope* that the minimum of the FCQS will also lead to a minimum of the modified objective function. Because of the large differences between the objective functions, this can be valid only in the case when all data are quite near to the quadric, because then the differences between the distance function are quite small. The combination of the modified distance with the determination of the prototypes corresponding to theorem 4.6 results in the modified fuzzy c-quadric shells algorithm [66]. (However, some large data sets caused convergence problems with the MFCQS.)

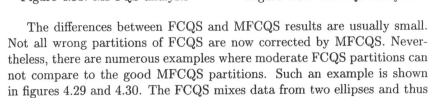

Figure 4.31: MFCQS analysis   Figure 4.32: MFCQS analysis

The differences between FCQS and MFCQS results are usually small. Not all wrong partitions of FCQS are now corrected by MFCQS. Nevertheless, there are numerous examples where moderate FCQS partitions can not compare to the good MFCQS partitions. Such an example is shown in figures 4.29 and 4.30. The FCQS mixes data from two ellipses and thus recognizes just one cluster correctly. The MFCQS, on the other hand, recognizes all three ellipses. The small cross with high memberships in the middle of the long-stretched ellipse shows that the distance function is only an approximation of the Euclidean distance – with Euclidean distance those pixel near the centre would have obtained low membership. (The two circle clusters have high memberships in their centres, as well; however, these can not be recognized very well in the figure, because these areas are on the respective contour of the other circle.) Strong deviations from results with Euclidean distance occur in the example shown in figure 4.32, especially at the two parabolas. These have a narrow region of high memberships, clearly outside the contour, as an elongation of the main axis, which does not happen with the FCQS. In some cases, areas like these can be responsible for a wrong partition made by MFCQS. The modified distance function provides Euclidean distances along the main axes of the quadric, but larger values remote from them. This way, the regions of higher memberships arise in the middle of the cluster.

The contours in figure 4.30 are thicker than in figure 4.29, because the value $\frac{1}{10}$ was assumed for the extension factors η in the possibilistic representation of both figures, and thus the contours get thinner because of the faster increasing distance with FCQS. In addition, the contour thickness of a FCQS cluster varies, whereas the thickness remains constant with MFCQS.

For figures 4.31 and 4.32, the same initialization and cluster parameters were chosen. The data sets only differ in the rotation angle of a single ellipse, however, MFCQS leads to completely different partitions. Phenomena like this occur for both FCQS and MFCQS, and it is difficult to predict if a data set will be partitioned well. If the initialization of the algorithms is not close enough to the intuitive partition, the quadric shells algorithms often get stuck in local minima because of the manifold cluster shapes. In such cases, even the modified distance function is no guarantee for a better result of the analysis. It can not compensate for the bad initialization.

Figure 4.33 shows a data set clustered by FCQS with $m = 4$. The contours were well recognized except the inner circle. The inner circle is hardly covered and the data vectors inside cannot shift the contour to the desired shape. If the MFCQS algorithm is applied to the result of FCQS, the modified distances and the resulting memberships cause the data of the inner circle to draw on the contour of the ellipse with a *greater force*. The result of the analysis is shown in figure 4.34; all clusters have been recognized correctly. If MFCQS is not initialized with the result of FCQS but with the initialization used for FCQS before, MFCQS does not recognize the intuitive partition. A substantial difference in these two cases is that in the first example there was a cluster with small distances for both FCQS

Figure 4.33: FCQS analysis   Figure 4.34: MFCQS analysis

and MFCQS, because the prototypes already approximated the data well. FCQS could be *cheated* with the resulting memberships, and provided the better partition. The influence of the modification is locally limited in this case and leads to the desired result. If there is not a good approximation, the distinctly different MFCQS distances are accordingly reflected in all memberships. In this case, FCQS less likely recognizes the correct contours with the modified memberships. MFCQS works well for *fine tuning* of FCQS partitions, but it often breaks down when the initialization is too far away from the correct partition. Then, the regions of high memberships close to the main axes of the MFCQS clusters further complicate the correct recognition, because they support the covering of the whole area. This can lead to a result like that in figure 4.35.

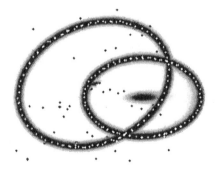

Figure 4.35: MFCQS analysis   Figure 4.36: P-MFCQS analysis

Figure 4.37: P-MFCQS analysis

There is also a possibilistic version of MFCQS. However, a *balanced proportion* of noise data to *noise-free* data is necessary. MFCQS minimizes the objective function for the data set in figure 4.37 by spending a part of a hyperbola for noise data only. The perfection with which the objective function is minimized here is remarkable; unfortunately this solution does not fit the ellipses. However, two extreme outliers are not the usual case, normally noise is scattered uniformly across the data set as in figure 4.36. Here, the 240 point data set contains 50 uniformly distributed noise data. The algorithm recognizes the ellipses well. However, possibilistic clustering becomes problematic when noise data occur in certain formations like point clouds. Then, the clusters adjust to the noise data again, similar to figure 4.37.

The modified distance causes a dramatic increase of the run time. With a data set of n data, that have to be partitioned into c clusters, $n \cdot c \cdot i$ fourth degree equations have to be solved for i iteration steps. For 300 data, 5 clusters, and 100 iteration steps, this means 150,000 (!) equations after all.

For the implementation, the function atan2 should be used for the determination of the modified distance, since the simple function atan produces imprecise results for values $k_1 - k_2$ close to zero because of the division by small values. Furthermore, the case of a scaled identity matrix has to be considered, especially when both k_3 and $k_1 - k_2$ are zero. If $k_3 = 0$, we can choose $\alpha = 0$ because we already deal with a diagonal matrix then. Floating point arithmetics should be used with double precision for the modified fuzzy c-quadric shells algorithm to avoid imprecisions in the computations in any case. If the computed distances are permanently distorted by round-off errors, convergence is impossible.

For higher dimensions, the modified distance leads to higher degree polynomials for which no explicit solution exists. In this case, a numerical technique is required again.

4.8 Computational effort

It should be pointed out once again that the algorithms were not optimized with respect to execution speed. We have the following time indices for FCS, AFCS, and FCES: $\tau_{FCS} = 6.58 \cdot 10^{-4}$, $\tau_{AFCS} = 2.24 \cdot 10^{-3}$, $\tau_{FCES} = 4.36 \cdot 10^{-3}$. As we can see from the numbers, finding a solution numerically is quite expensive. The large difference between FCS and AFCS/FCES reflects the dimension of the equation systems that have to be solved. Essentially, shorter execution times cannot be expected from other iteration procedures for the solution of the equation systems, since these algorithms are mostly computationally even more complicated. (With the Levenberg-Marquardt algorithm, in each iteration step a further iteration is performed leading to three recursive iteration procedures.)

Algorithms which determine the prototypes directly are much faster: $\tau_{FCSS} = 1.03 \cdot 10^{-4}$, $\tau_{FCE} = 2.40 \cdot 10^{-4}$, $\tau_{FCQS} = 1.58 \cdot 10^{-4}$. MFCQS is much slower compared with FCQS, because many fourth degree polynomials have to be solved: $\tau_{MFCQS} = 1.12 \cdot 10^{-3}$.

The data sets contained between 62 and 315 data, and needed (apart from few exceptions) usually much less than 100 iteration steps. The initialization procedures are described in the text. Several algorithms for the recognition of ellipses are compared in [36].

Chapter 5

Polygonal Object Boundaries

Before we take a closer look at the ideas leading to fuzzy clustering algorithms for detecting rectangles, and later on more general polygons [43], we briefly review the previously described algorithms. All the fuzzy clustering techniques discussed in chapters 2, 3, and 4 are based on the same underlying principal of algorithm 1. Nevertheless, their ability to detect the type of clusters they are designed for, is not always comparable. These differences appear between algorithms of the same class as well as of algorithms of different classes (solid, linear and shell clustering). The reasons for this are first, the increasing number of degrees of freedom within each class, but also the more complicated cluster shapes of the higher classes. The number of local minima of the objective function, and therefore also the probability to converge in an undesired local minimum is increasing with the complexity of the cluster shapes. The problems caused by this tendency are more serious for *very flexible* cluster shapes. Two effects can be observed:

- The influence of the initialization increases within the same class of clustering algorithms. The simple fuzzy c-means algorithm is more or less independent of the initialization and almost always converges with the same result, whereas the more complicated Gath-Geva algorithm often tends to converge in a result not too far away from the initial prototypes. Therefore, varying the initialization usually leads to different results. A similar relation can be observed between the fuzzy c-shells and the fuzzy c-quadric shells algorithm. The best results are often obtained for a sequence of algorithms where each

115

more complicated algorithm is initialized with the clustering result computed by its simpler predecessor algorithm which allows a little less freedom in comparison to its successor. The sequence fuzzy c-means, Gustafson-Kessel algorithm, and Gath-Geva algorithm is almost ideal, since, one after the other, position, shape, and size of the clusters are incorporated in the corresponding algorithm. For shell clustering a useful sequence is fuzzy c-shells, fuzzy c-ellipses, and the modified fuzzy c-quadric shells algorithm.

- For the higher classes of clustering algorithms, small spatial distances of the data (among themselves) become less important for the assignment of the data to the same cluster. Although the fuzzy c-means and the Gath-Geva algorithm are based on completely different distance measures, the (Euclidean) distance of a datum to the cluster centre still has a strong influence on the assignment of the datum to the cluster. The distance measures on which these algorithms are based always increase with the Euclidean distance, possibly with different speeds for varying directions. For the fuzzy c-varieties algorithm there is one (or more direction) for which the algebraic distance between datum and cluster centre does not increase at all with the Euclidean distance. And for shell clustering algorithms the algebraic distance (starting at the cluster centre) usually first decreases until we reach the cluster contour and then increases again. In this way, data that are not closely connected can be gathered within one cluster. An example is illustrated in figure 4.4, where one circle shell is only covered partly by data in two different regions. The additional condition that the contour of a circle or ellipse should be marked completely by data vectors is hard to formalize. The partition of figure 4.4 contradicts intuition, because the proximity of neighbouring data vectors is desired. However, this cannot be seen from the distance function. Without this constraint we are able to detect ellipses which are only partly covered by data, as in figure 4.8. But the clustering result does not tell us whether we obtained the computed ellipses from small segments or full ellipses (cf. also section 7.2.3). Requiring that the data are closely connected, the corresponding cluster shape would no longer be ellipses, but segments of ellipses. Thus a result as in figure 4.37 would not be admissible. On the other hand, without this constraint, for some data sets, shell clustering algorithms tend to cover the data by large shells without detecting all intuitively visible ellipses.

In particular, cluster shapes that extend to infinity like lines, parabolas, or hyperbolas tend to grab data near their contour that do actually not belong to the cluster. Unfortunately, these incorrectly assigned data with their high membership degree sometimes prevent the cluster from adapting to the optimal parameters. In such cases, limiting the cluster to regions where data have a small spatial distance could be helpful.

Sometimes, this problem can be overcome using the heuristically modified distance measure, as already known from the fuzzy c-elliptotypes algorithm. In the case of shell clusters k, a cluster centre z may be introduced, to which the Euclidean distance is incorporated into the new distance measure:

$$d_{modified}(x, k) = \alpha \|z - x\| + (1 - \alpha) d_{shell\ distance}(x, k) \ .$$

The centre may be updated using the shell clustering memberships and the prototype update of the fuzzy c-means algorithm. Switching to this modified distance function after converging, some misclassifications can be corrected. The value of α might be chosen to be about 10^{-4}.

The capabilities of a clustering algorithm do not only depend on the cluster shape, but also very much on the underlying distance function. For algorithms that allow for a great variety of cluster shapes, an almost Euclidean distance function is very important for good performance in the sense of detecting clusters correctly and – especially for shell clustering – to get improved results by a small increase of the fuzzifier. Larger fuzzifiers tend to smoothen the distance function so that undesired local minima that are not too extreme can be avoided by this technique. However, many examples make us believe that for highly non-Euclidean distance functions, the smoothing effect is not strong enough to help in significantly avoiding local minima. In fact, for a non-Euclidean distance function the global optimum might be totally different from the intuitive partition, and this is especially true when we have to deal with some outliers or noise. This again emphasizes the importance of the Euclidean distance function that is mainly responsible for the cluster partition from the human point of view.

5.1 Detection of rectangles

In the following we examine whether the basic clustering algorithm is suitable for detecting clusters in the form of rectangular shells. In principle, linear clustering algorithms are capable of finding rectangles as some of the figures in section 3 demonstrate. For n rectangles in an image, at most $4 \cdot n$ clusters in the form of lines are needed. However, for a higher number of clusters undesired results that just cover the data by a large number of parallel lines may occur (see figure 3.4). In addition, the linear clusters influence each other according to their infinite extension. Finally, for detecting rectangles it is not sufficient to have just a number of lines, but another algorithm is needed to find groups of four lines that form rectangles.

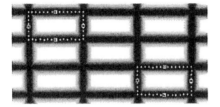   

Figure 5.1: AFC analysis   Figure 5.2: AFC analysis

For this task it is necessary to know the direction of the lines, as well as their end-points. Although the lines marking the two rectangles in figures 5.1 and 5.2 are identical, the edges of the corresponding rectangles are not. The cluster parameters determining the centres of the linear clusters can help to find the rectangular groups. However, these centres are not necessarily the mid-points of the edges, when the data are inhomogeneously spread on the edge. A preselection of potential rectangular groups can be obtained in this way, but the cluster centres are not suitable for an optimal adaptation of the rectangles to the data, since the orthogonality is ignored. Thus an algorithm that is especially tailored for rectangles may be better than just detecting lines first and finding rectangular group in a second step. For the development of such an algorithm for rectangular clusters we should take into account the considerations on the cluster shape and the distance function we discussed at the beginning of this chapter. Of course, an explicit computation scheme for the prototypes given the membership degrees is desirable in order to design a reasonably fast algorithm. After the design of such an algorithm, we have to develop suitable validity measures to be able to determine the number of clusters (rectangles) in a given image. We will not discuss the problem here, since these validity measures can be defined analogously to those described in chapter 7. In particular, in opposition to general quadrics, the distance vector and the distance of a datum to a given line or rectangle can be computed easily (see also section 5.2). Thus the global validity measures from section 7.1.2 can be directly applied. Refraining from taking the mean over all clusters, we can even derive local validity measures for rectangles from the global ones. Therefore, we restrict ourselves in the following to the shell clustering algorithms for rectangles, without discussing the validity measures here.

The cluster form *rectangle* is given, but we are still free to choose a suitable representation. A rectangular contour cannot be defined in an analytical form as easily as circles or quadrics. Therefore, it might be reasonable to accept a compromise that either tolerates deviations from the exact cluster shapes for the sake of a faster algorithm or accepts high computational efforts. With these considerations in mind, we take a closer look at the possibilities and restrictions for designing a clustering algorithm for rectangles.

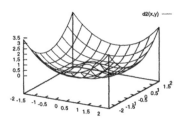
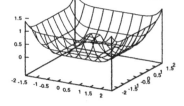

Figure 5.3: $p = 2$ Figure 5.4: $p = 8$

We restrict ourselves to the case of a two-dimensional data space $D :=$ $\mathbb{R}^2$. We have to define a distance function yielding the value zero on a rectangle. For the moment we do not consider a rotation of the rectangle. A square can be approximated on the basis of a p-norm $|| \cdot ||_p : \mathbb{R}^n \to$ $\mathbb{R}_{\geq 0}$, $(x_1, x_2, \ldots, x_n) \mapsto \sqrt[p]{\sum_{i=1}^n |x_i|^p}$. For $p = 2$ we obtain the Euclidean norm. Using $d_p : \mathbb{R}^2 \times \mathbb{R}^2 \to \mathbb{R}$, $(x, v) \mapsto (||x - v||_p - 1)^2$ with $p = 2$ we obtained a circle with radius 1 for the fuzzy c-shells algorithm. Increasing the value of p the circle for $p = 2$ changes more and more to a square, as can be seen in figures 5.3 and 5.4 showing the function d_p for $p = 2$ and $p = 8$. Looking at these figures, one can imagine the continuous transformation of the circle to a square while increasing p from 2 to ∞. Replacing the constant value 1 by an edge length $k \in \mathbb{R}$ in the function d_p, squares of different sizes can be described. For rectangles whose edge lengths are not identical, as in the case of squares, we need two parameters, scaling the distances in the directions of the coordinate axes. Describing an axes-parallel rectangle by the parameters $(v_0, v_1, r_0, r_1) \in \mathbb{R}^4$ with its centre (v_0, v_1) and the half edge lengths r_0 and r_1, the following remark provides a suitable distance function.

Remark 5.1 (Prototypes for rectangles using p-norms) *Let $p \in \mathbb{N}$, $D := \mathbb{R}^2$, $X = \{x_1, x_2, \ldots, x_n\} \subseteq D$, $C := \mathbb{R}^4$, $c \in \mathbb{N}$, $R := \mathcal{P}_c(C)$, J corresponding to (1.7) with $m \in \mathbb{R}_{>1}$ and*

$$d^2 : D \times C \quad \to \quad \mathbb{R}_{\geq 0},$$

$$((x_0, x_1), (v_0, v_1, r_0, r_1)) \quad \mapsto \quad \left(\sqrt[p]{\sum_{s=1}^{2} \left| \frac{x_s - v_s}{r_s} \right|^p} - 1 \right)^2.$$

If J is minimized with respect to all probabilistic cluster partitions $X \to F(K)$ with $K = \{k_1, k_2, \ldots, k_c\} \in R$ and given membership degrees $f(x_j)(k_i) = u_{i,j}$ by $f : X \to F(K)$, then we have for $k_i = (v_{i,0}, v_{i,1}, r_{i,0}, r_{i,1})$, $q \in \{p, p-1\}$, and $s \in \{0,1\}$:

$$0 = \sum_{j=1}^{n} u_{i,j}^m \frac{\sqrt[p]{d_{i,j}} - 1}{\sqrt[p]{d_{i,j}^{p-1}}} \left(\frac{x_{j,s} - v_{i,s}}{r_{i,s}} \right)^q ,$$

where $d_{i,j} = \left(\frac{x_{j,0} - v_{i,0}}{r_{i,0}} \right)^p + \left(\frac{x_{j,1} - v_{i,1}}{r_{i,1}} \right)^p$.

Proof: The probabilistic cluster partition $f : X \to F(K)$ shall minimize the objective function J. Then all partial derivatives of J have to be zero. With $k_i = (v_{i,0}, v_{i,1}, r_{i,0}, r_{i,1}) \in K$ and $s \in \{0,1\}$ we obtain:

$$0 = \frac{\partial J}{\partial v_{i,s}} = \sum_{j=1}^{n} -2u_{i,j}^m (\sqrt[p]{d_{i,j}} - 1) \frac{1}{p} d_{i,j}^{\frac{1-p}{p}} p \left(\frac{x_{j,s} - v_{i,s}}{r_{i,s}} \right)^{p-1} \frac{1}{r_{i,s}}$$

$$\Leftrightarrow \quad 0 = \sum_{j=1}^{n} u_{i,j}^m \frac{\sqrt[p]{d_{i,j}} - 1}{\sqrt[p]{d_{i,j}^{p-1}}} \left(\frac{x_{j,s} - v_{i,s}}{r_{i,s}} \right)^{p-1} \quad \text{and}$$

$$0 = \frac{\partial b}{\partial r_{i,s}} \quad \Leftrightarrow \quad 0 = \sum_{j=1}^{n} u_{i,j}^m \frac{\sqrt[p]{d_{i,j}} - 1}{\sqrt[p]{d_{i,j}^{p-1}}} \left(\frac{x_{j,s} - v_{i,s}}{r_{i,s}} \right)^{p} .$$

■

There is no analytical solution for this system of non-linear equations for larger values of p, since $(x_{j,0} - v_{i,0})$ and $(x_{j,1} - v_{i,1})$ appear to the power of p and $(p-1)$. Thus we can again obtain an approximate solution by applying a numerical algorithm.

The scaling factors r_0 and r_1 for taking different edge lengths of the rectangle into account cause a deviation from the Euclidean distance, since data points with the same Euclidean distance to the rectangle may yield different values for d. The edge lengths of the rectangle are proportional to r_0 and r_1 so that the distances along the axes are more shortened for larger than for smaller rectangles. Different edge lengths also lead to different scalings of the distance function in the x- and y-direction. The distance within a rectangle will never exceed the value 1, regardless of the size of the rectangle. Thus we are very far away from a Euclidean distance measure, and have to cope with similar problems as in the case of the AFCS and the FCQS algorithms.

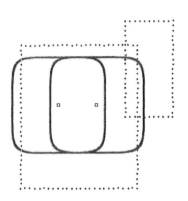

Figure 5.5: p-norm analysis

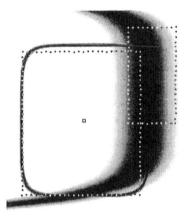

Figure 5.6: p-norm analysis

In this simple form the algorithm does not yield satisfactory results. Figures 5.5 and 5.6 show a data set containing two rectangles; figure 5.5 illustrates the initialization, figure 5.6 the result after one iteration step. The left initialization cluster already approximates the larger rectangle after one iteration step quite well. The other cluster has to minimize the distances to the data of the smaller rectangle and the right part of the larger rectangle. The Newton iteration tries to find an optimum by a gradient descent technique. When the centre of the remaining cluster is shifted to the right, then the horizontal edge length has to be shortened also in order to minimize the distances to data at the right side. However, this shortening enforces an enlargement of the distances along this axis so that even the same location of the right edge would lead to larger distances. The Newton iteration will therefore not carry out these changes, since it would have to pass a local maximum. Moving the right centre point in the opposite direction as desired and stretching the right edge at the same time, the distances are shortened for the same reason. In each iteration step the Newton iteration stretches the horizontal edge even more, until

the distance of the data on the right edge is almost zero and the Newton
iteration seems to converge. In the following steps the Newton iteration
converges again and again towards this local minimum. Finally the clustering algorithm terminates with the undesired result in figure 5.6, since
for a very large edge length, the membership degrees remain more or less
stable. We can also see that for a large edge length the approximation of
the rectangle on the basis of a p-norm is no longer acceptable.

Summarizing these considerations we can say:

- A numerical solution of the system of non-linear equations may result
  in the convergence in a local minimum with an almost infinite edge
  length. A constraint for the edge lengths avoiding such problems is
  almost impossible without restricting the possible rectangular shapes.

- In contrast to circular clusters, a single segment is not sufficient for
  determining the cluster parameters. In principle, three data points
  are already enough to find a circle approximating them. If the data
  points of at least one edge are completely missing in a rectangle,
  there are infinitely many possible solutions. The intuitive solution
  – the smallest rectangle fitting to the corresponding data – is even
  counteracted by the distance function in remark 5.1.

- For larger rectangles and smaller values of p, the corners of the rectangles are not well approximated by the clusters (see also page 132).

- The convergence to solutions with (almost) infinite edge lengths is
  not caused by the algorithm itself, but by the procedure to solve
  the system of non-linear equations. The solution determined by the
  Newton iteration is actually not a proper solution, since only arbitrary
  small values, but not the value zero are assumed.

The last observation shows that the algorithm becomes useless only in
connection with a numerical procedure like the Newton iteration. With
a better initialization, even the Newton iteration is able to find a better
cluster partition. Indeed, choosing

$$v_i^\top = (v_{i,0}, v_{i,1})^\top = \frac{\sum_{j=1}^n u_{i,j}^m x_j}{\sum_{j=1}^n u_{i,j}^m},$$

$$(r_{i,0}, r_{i,1})^\top = \frac{\sum_{j=1}^n u_{i,j}^m (x_j - v_i)}{\sum_{j=1}^n u_{i,j}^m}$$

as the initial values for the ith cluster, $i \in \mathbb{N}_{\leq c}$, in figure 5.5 leads to the
desired clusters. In this case, the centre of each rectangle is computed as

the centre of gravity of the data in the corresponding rectangle before the
Newton iteration is started. The above initializations of the edge lengths
ignore to which edge a datum actually belongs. Nevertheless, these values
provide a good heuristic initialization for the Newton iteration.

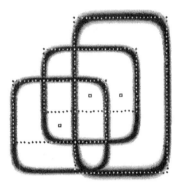

Figure 5.7: p-norm analysis

Despite the success of the algorithm after the heuristic initialization of
the Newton iteration, the algorithm still remains a very rough approach.
The contour induced by the distance function only approximates rectangles. For simple images this might be sufficient, but in general problems
as in figure 5.7 appear, where the bad approximation of the data on the
horizontal edges, especially in the corners, prevents the algorithm from detecting the rectangles correctly. This bad approximation leads to smaller
membership degrees so that other rectangles tend to cover these data as
well. For instance, the lower left cluster approximates one corner of the
large rectangle on the right instead of concentrating on its correct edge.
Even the sparsely covered data on this edge do not force the cluster into
the right shape due to the distance function that is bounded by one in the
interior of the rectangle (independently of the size of the rectangle).

In some cases, when the Newton iteration computes long edges and these
values are reset to much smaller ones by the heuristic Newton initialization,
the algorithm does not converge at all.

As mentioned before, in the special case of squares the scaling of the
axes in the distance function can be avoided by defining $d_p : \mathbb{R}^2 \times \mathbb{R}^2 \to$
$\mathbb{R}$, $(x, v) \mapsto (\|x - v\|_p - r)^2$. Then the distance of a datum to the square
centre is independent of the direction. The parameter r determines the
length of the square's edges in the same way as it determines the radius of
a circle in the case of FCS. Here, an infinite edge length does not lead to a

supposed zero passage, and the Newton iteration can be initialized by the last iterations result. The procedure is therefore applicable to the detection of axes-parallel squares.

Introducing an additional parameter for a rotation, the algorithm could be generalized to arbitrary rectangles. We would have to replace the expressions $(x_{j,0} - v_{i,0})$ and $(x_{j,1} - v_{i,1})$ by $(x_{j,0} - v_{i,0})\cos(\alpha) - (x_{j,1} - v_{i,1})\sin(\alpha)$ and $(x_{j,0} - v_{i,0})\sin(\alpha) + (x_{j,1} - v_{i,1})\cos(\alpha)$, respectively. Since the axes-parallel version of the algorithm is already not very successful, we do not consider this generalization here.

For a more precise description of the contour of a rectangle, we need a different distance function. We would like to avoid a scaling of the axes and would prefer to incorporate varying edge lengths by subtraction, as in the case of the square, instead of by division. Since we have two pairs of parallel edges in a rectangle, we have to introduce the subtraction for both axes separately. Figure 5.8 shows the graph of the function

$$\mathbb{R}^2 \to \mathbb{R}, \quad (x,y) \mapsto \left((x^2 - 1)(y^2 - 1)\right)^2.$$

The factors $(x^2 - 1)$ and $(y^2 - 1)$ become zero for $x, y \in \{-1, 1\}$. Thus the set of points where the function assumes the values zero are two pairs of parallel lines that meet in the corners of a square. However, these lines continue to infinity and do not end in the corners of the square, so that we will encounter the same problems as in the case of linear clustering and line segments.

Remark 5.2 (Prototypes for rectangles using four lines) *Let $D :=$ $\mathbb{R}^2$, $X = \{x_1, x_2, \ldots, x_n\} \subseteq D$, $C := \mathbb{R}^4$, $c \in \mathbb{N}$, $R := \mathcal{P}_c(C)$, J corresponding to (1.7) with $m \in \mathbb{R}_{>1}$ and*

$$d^2 : \mathbb{R}^2 \times C \to \mathbb{R}_{\geq 0}, \quad ((x_0, x_1), (v_0, v_1, r_0, r_1)) \mapsto \left(\prod_{s=0}^{1}(x_s - v_s)^2 - r_s\right)^2.$$

If J is minimized with respect to to all probabilistic cluster partitions $X \to F(K)$ with $K = \{k_1, k_2, \ldots, k_c\} \in R$ and given membership degrees $f(x_j)(r_i) = u_{i,j}$ by $f : X \to F(K)$, then we have for $k_i = (v_{i,0}, v_{i,1}, r_{i,0}, r_{i,1})$:

$$0 = \sum_{j=1}^{n} u_{i,j}^m ((x_{j,0} - v_{i,0})^2 - r_{i,0})\left((x_{j,1} - v_{i,1})^2 - r_{i,1}\right)^2 (x_{j,0} - v_{i,0})$$

$$0 = \sum_{j=1}^{n} u_{i,j}^m ((x_{j,0} - v_{i,0})^2 - r_{i,0})^2 \left((x_{j,1} - v_{i,1})^2 - r_{i,1}\right) (x_{j,1} - v_{i,0})$$

$$0 \; = \; \sum_{j=1}^{n} u_{i,j}^{m} ((x_{j,0} - v_{i,0})^2 - r_{i,0}) \, ((x_{j,1} - v_{i,1})^2 - r_{i,1})^2$$

$$0 \; = \; \sum_{j=1}^{n} u_{i,j}^{m} ((x_{j,0} - v_{i,0})^2 - r_{i,0})^2 \, ((x_{j,1} - v_{i,1})^2 - r_{i,1}).$$

Proof: The equations are obtained directly by requiring that the partial derivatives of the objective function have to be zero. ∎

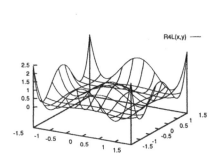

Figure 5.8: Distance function for approximation of rectangles by four lines

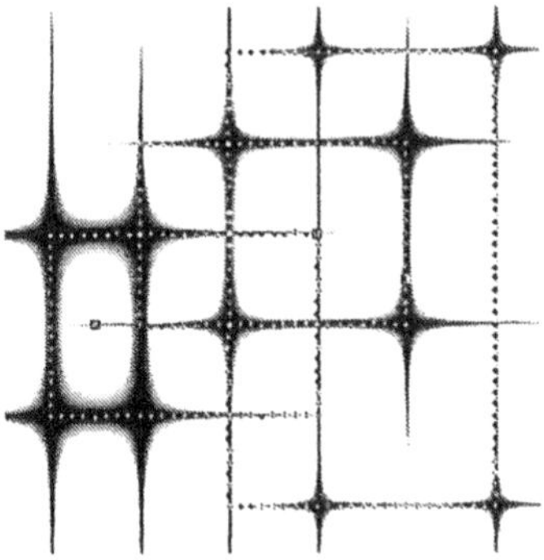

Figure 5.9: Clustering result

One disadvantage of this distance function is that the clusters are essentially not rectangles but four lines. In this way the data in figure 5.9 are covered quite well by the clusters, although the rectangles do not correspond to the expected result (for a better view of the data set, see also figure 5.11). This algorithm just combines sets of four lines to a rectangle without improving much in comparison to a linear clustering algorithm, except that orthogonality of the edges is guaranteed. The considerations and examples in the chapter on linear clustering show that it is important for the distance function (and the membership degrees) to increase (respectively decrease) outside the cluster contour. If we wanted to further elaborate this approach, we would have to incorporate the Euclidean distance to the centre of the rectangle in the same way as for the fuzzy c-elliptotypes or the adaptive fuzzy clustering algorithm , in order to avoid clusters extending to infinity.

The extension of the lines further than the considered rectangle is caused by the multiplication of the expressions $((x_{j,0} - v_{i,0})^2 - r_{i,0})$ and $((x_{j,1} - $

$v_{i,1})^2 - r_{i,1})$. Figure 5.10 shows the graph of the function

$$\mathbb{R}^2 \to \mathbb{R}, \quad (x, y) \mapsto (x^2 - 1)^2 + (y^2 - 1)^2.$$

The corners of the rectangle have the minimum distance for this distance function, along the edges the distance is a little bit higher but lower than outside the rectangle contour.

Remark 5.3 (Prototypes for rectangles using four points) *Let $D := \mathbb{R}^2$, $X = \{x_1, x_2, \ldots, x_n\} \subseteq D$, $C := \mathbb{R}^4$, $c \in \mathbb{N}$, $R := \mathcal{P}_c(C)$, J corresponding to (1.7) with $m \in \mathbb{R}_{>1}$ and*

$$d^2 : \mathbb{R}^2 \times C \to \mathbb{R}_{\geq 0}, \ ((x_0, x_1), (v_0, v_1, r_0, r_1)) \mapsto \sum_{s=0}^{1}((x_s - v_s)^2 - r_s)^2.$$

If J is minimized with respect to all probabilistic cluster partitions $X \to F(K)$ with $K = \{k_1, k_2, \ldots, k_c\} \in R$ and given membership degrees $f(x_j)(k_i) = u_{i,j}$ by $f : X \to F(K)$, then we have for $k_i = (v_{i,0}, v_{i,1}, r_{i,0}, r_{i,1})$ and $s \in \{0, 1\}$:

$$v_{i,s} = \frac{S_{i,3} - \frac{S_{i,1}S_{i,2}}{S_{i,0}}}{2\left(S_{i,2} - \frac{S_{i,1}^2}{S_{i,0}}\right)}, \tag{5.1}$$

$$r_{i,s} = \frac{S_{i,2} - 2v_{i,s}S_{i,1}}{S_{i,0}} + v_{i,s}^2, \tag{5.2}$$

where $S_{i,l} = \sum_{j=1}^{n} u_{i,j}^m x_{j,s}^l$ for $l \in \mathbb{N}_{<4}^*$.

Proof: Let the probabilistic cluster partition $f : X \to F(K)$ minimize the objective function J. Again, the partial derivatives have to be zero. With $k_i = (v_{i,0}, v_{i,1}, r_{i,0}, r_{i,1}) \in K$ and $s \in \{0, 1\}$ we have:

$$\frac{\partial J}{\partial v_{i,s}} = 0 = -4\sum_{j=1}^{n} u_{i,j}^m((x_{j,s} - v_{i,s})^2 - r_{i,s})(x_{j,s} - v_{i,s}), \tag{5.3}$$

$$\frac{\partial J}{\partial r_{i,s}} = 0 = -\sum_{j=1}^{n} u_{i,j}^m((x_{j,s} - v_{i,s})^2 - r_{i,s}).$$

From the latter equation we can derive the formula (5.2) for the parameter r. Rearranging equation (5.3), we obtain, together with (5.2):

$$0 = \sum_{j=1}^{n} u_{i,j}^m \left[x_{j,s}^3 - 2v_{i,s}x_{j,s} + v_{i,s}^2 - r_{i,s})(x_{j,s} - v_{i,s})\right]$$

$$= \sum_{j=1}^{n} u_{i,j}^m \left[v_{i,s}^3 (-1+1) + v_{i,s}^2 (x_{j,s} - x_{j,s} + 2x_{j,s} - \frac{2S_{i,1}}{S_{i,0}}) \right.$$

$$\left. + v_{i,s}(-2x_{j,s}^2 + x_{j,s}\frac{2S_{i,1}}{S_{i,0}} - x_{j,s} + \frac{S_{i,2}}{S_{i,0}}) + x_{j,s}^3 - x_{j,s}\frac{S_{i,2}}{S_{i,0}} \right]$$

$$= v_{i,s}^2 (2S_{i,1} - S_{i,0}\frac{2S_{i,1}}{S_{i,0}}) + v_{i,s}(-3S_{i,2} + 2S_{i,1}\frac{S_{i,1}}{S_{i,0}} + \frac{S_{i,2}}{S_{i,0}})$$

$$+ S_{i,3} - S_{i,1}\frac{S_{i,2}}{S_{i,0}}$$

$$= v_{i,s}(-2S_{i,2} + \frac{2S_{i,1}^2}{S_{i,0}}) + S_{i,3} - S_{i,1}\frac{S_{i,2}}{S_{i,0}}.$$

From this equation we can derive the formula (5.1) for the parameter $v_{i,s}$. Taking this formula for $v_{i,s}$ into account, we can finally compute $r_{i,s}$ on the basis of (5.2). ■

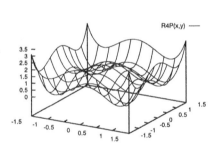

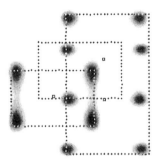

Figure 5.10: Distance function for approximation of rectangles by four points

Figure 5.11: Clustering result

Changing the distance function from a product to a sum, we are able to provide an explicit solution for the prototypes. In the same way that the cluster shape in the product resembles four lines, we obtain for the sum four point-like clusters as can be seen in figure 5.11. (The small squares indicate the centres of the rectangles that are induced by a collection of four regions of high membership degrees. The points on the right side from top to bottom are alternatingly assigned to two different clusters.) The corners of the clusters do not fit completely to the corners of the rectangles, but behave in the same manner as fuzzy c-means clusters. The distance is minimized when the cluster corners are placed a little bit inside the corners of the rectangles. Therefore, the correct rectangle parameters cannot be computed with this algorithm. Moreover, edges of some rectangles may be covered insufficiently. As already mentioned, the distance function on the edges is increasing in the direction to the edge centres. So data points near the edge centres may have small membership degrees and can attract corners of other clusters, resulting in a counter-intuitive clustering result. Another disadvantage of this approach is that the distance function is not tailored for closely connected data on the contour, but prefers the four corners. However, four point-like clusters can be fitted to shapes other than rectangles as well.

So far, the proposed rectangle algorithms do not yield completely satisfactory results, because the cluster form induced by the distance function is not an exact rectangle, but only an approximation of it. We can expect better results from the algorithms

- if the *rounded* corners for the p-norm are compensated by higher values for p. As the limit $p \to \infty$ we obtain the maximum norm. (See also page 132.)

- if the lines are bounded by the rectangles.

- if instead of four point-like clusters four corner clusters describe the four corners of the rectangle.

These improvements require the use of absolute values, respectively the minimum or maximum function, that can be expressed in terms of absolute values. Figure 5.12 shows the graph of the function $d_{max} : \mathbb{R}^2 \to \mathbb{R}$, $(x, y) \mapsto (max(|x|, |y|) - 1)^2$. The expression $max(|x|, |y|)$ is obtained as the limit for $p \to \infty$ in the formula for the p-norm of (x, y). Figure 5.13 illustrates the graph of the function $d_{min} : \mathbb{R}^2 \to \mathbb{R}$, $(x, y) \mapsto min^2(x, y)$, which can be taken as a distance function to detect corners. The edges of the corners extend to infinity as the linear clusters in the fuzzy c-varieties algorithm. The function $d_{edge} : \mathbb{R}^2 \to \mathbb{R}$, $(x, y) \mapsto min^2(x, y) + max(|x|, |y|)$ takes also the distance to the corner in the direction of the edges into account, but less than in other directions, since we want to detect a corner consisting of two edges and not a single point. The graph of this function is shown in figure 5.14.

Unfortunately, the appearance of the maximum or minimum function in the distance function causes new problems. Writing these functions in the form $max(x, y) = \frac{x+y+|x-y|}{2}$ and $min(x, y) = \frac{x+y-|x-y|}{2}$, we need the absolute value which is not differentiable in zero. We have avoided this problem until now by taking the square. But when we take the square of the

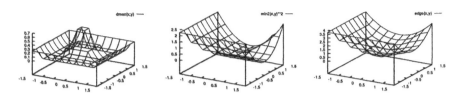

Figure 5.12: d_{max} Figure 5.13: d_{min} Figure 5.14: d_{edge}

maximum or minimum function, the absolute value will not be eliminated. The derivative of the objective function cannot be computed everywhere. The derivative of the function $|f(x)|$, $f : \mathbb{R} \to \mathbb{R}$, is $f'(x)\,\mathrm{sign}(f(x))$ for all $x \in \mathbb{R}\backslash\{0\}$. If f is the distance function, then it is not differentiable right on the cluster contour. For the fuzzy c-shells algorithm the Euclidean norm appears non-squared so that the same problems appear as for the absolute value function. However, the problematic values are not on the cluster contour, but further away. (In the resulting equation a division by $\|x - v\|$ appears so that the critical case occurs only if a datum is in the centre of the corresponding circle. But such a datum is not typical for the circle contour and does not influence the computation of the correct circle parameters based on the relevant data points, as long as we avoid the division by zero.)

Even more importantly is that the absolute value still appears in the derivative. The fuzzy c-shells algorithm needs a numerical procedure to solve a system of non-linear equations to compute the prototypes. An explicit solution for the prototypes cannot usually be found, when prototype parameters appear within absolute values, so that the equations can not be solved with respect to these parameters. A computational efficient algorithm for determining the prototypes needs explicit solutions for the parameters. Obviously, the price for the exact rectangular cluster shape on the basis of the maximum or minimum function in the distance function is a non-analytical solution for the prototype parameters. A modification of the distance function in remark 5.3 to

$$d : \mathbb{R}^2 \times \mathbb{R}^4 \to \mathbb{R}_{\geq 0}, \ ((x_0, x_1), (v_0, v_1, r_0, r_1)) \mapsto \max_{s \in \{0,1\}} \left((x_s - v_s)^2 - r_s\right)^2$$

leads, for the derivative of the objective function, to

$$0 = \frac{\partial b}{\partial v_s}$$

$$= \sum_{j=1}^{n} \left\{ \begin{array}{ll} -4((x_{j,s} - v_s)^2 - r_s)(x_{j,s} - v_s) & \text{if } j \in S \\ 0 & \text{if } j \notin S \end{array} \right\}$$

$$= -4 \sum_{j \in S} ((x_{j,s} - v_s)^2 - r_s)(x_{j,s} - v_s)$$

for $s \in \{0,1\}$, $\tilde{s} = 1 - s$ and $S := \{j \in \mathbb{N}_{\leq n} \mid ((x_{j,s} - v_s)^2 - r_s)^2 \geq ((x_{j,\tilde{s}} - v_{\tilde{s}})^2 - r_{\tilde{s}})^2\}$.

This partial derivative is similar to the partial derivative (5.3). The only difference lies in the terms of the sums. In (5.3) there is one term for each datum, now only the data in the set S have to be considered. In remark 5.3 this equation was necessary for determining an explicit formula for v_s. Without having v_s we are unable to determine the set S and do not know which terms appear in the sum. However, approximating the prototypes on the basis of a Newton iteration, we are able to design a clustering algorithm for rectangles, since the objective function has to be evaluated for certain instances only. In this case, the set S as well as the parameters for the rectangles are known. When we are satisfied with an iterative numerical procedure to determine the prototypes, cluster algorithms for detecting rectangles can be designed in the way described above. Regarding the distance function, the following theorem describes a very promising approach.

Theorem 5.4 (Prototypes for rectangles using the max-norm)
Let $D := \mathbb{R}^2$, $X = \{x_1, x_2, \ldots, x_n\} \subseteq D$, $C := \mathbb{R}^2 \times \mathbb{R}^2 \times \mathbb{R}$, $c \in \mathbb{N}$, $R := \mathcal{P}_c(C)$, J corresponding to (1.7) with $m \in \mathbb{R}_{>1}$ and

$$d^2 : \mathbb{R}^2 \times C \quad \rightarrow \quad \mathbb{R}_{\geq 0},$$

$$(x, (v, r, \varphi)) \quad \mapsto \quad \max \left\{ \left| (x - v)^\top \begin{pmatrix} \cos(\varphi + \frac{s\pi}{2}) \\ \sin(\varphi + \frac{s\pi}{2}) \end{pmatrix} \right| - r_s \;\middle|\; s \in \{0,1\} \right\}^2 .$$

If J is minimized with respect to all probabilistic cluster partitions $X \rightarrow F(K)$ with $K = \{k_1, k_2, \ldots, k_c\} \in R$ for given membership degrees $f(x_j)(k_i) = u_{i,j}$ by $f : X \rightarrow F(K)$, then we have for $k_i = (v, r, \varphi)$, $n_s = \begin{pmatrix} \cos(\varphi + \frac{s\pi}{2}) \\ \sin(\varphi + \frac{s\pi}{2}) \end{pmatrix}$ (where $s \in \{0,1\}$) and $X_0 = \{j \in \mathbb{N}_{\leq n} \mid (x_j - v)^\top n_0 - r_0 > (x_j - v)^\top n_1 - r_1\}$, $X_1 = \mathbb{N}_{\leq n} \backslash X_0$:

$$\frac{\partial J}{\partial v} = 0 \Rightarrow 0 = \sum_{j \in X_s} u_{i,j}^m (|(x_j - v)^\top n_s| - r_s) \operatorname{sign}((x_j - v)^\top n_s),$$

$$\frac{\partial J}{\partial r_s} = 0 \Rightarrow 0 = \sum_{j \in X_s} u_{i,j}^m (|(x_j - v)^\top n_s| - r_s),$$

for $s \in \{0,1\}$, and

$$\frac{\partial J}{\partial \varphi} = 0 \Rightarrow$$

$$0 = \sum_{j \in X_0} u_{i,j}^m (|(x_j - v)^\top n_0| - r_0) \, \mathrm{sign}((x_j - v)^\top n_0)(x_j - v)^\top n_1$$

$$+ \sum_{j \in X_1} u_{i,j}^m - (|(x_j - v)^\top n_1| - r_1) \, \mathrm{sign}((x_j - v)^\top n_1)(x_j - v)^\top n_0.$$

Proof: The equations follow directly from the partial derivatives. ∎

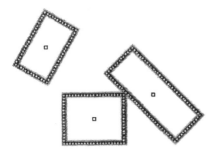

Figure 5.15: Maximum norm analysis   Figure 5.16: Maximum norm analysis

On the basis of a reasonable initialization (ten fuzzy c-means steps and ten fuzzy c-shells steps) this algorithm provides the desired results for the data sets in figures 5.5 and 5.11. One reason for the good results is the distance function, that computes the Euclidean distance from the cluster contour. The absolute value of the scalar product within the maximum computes the distance of datum x_j to the cluster centre v along the axes of a coordinate system that is rotated by the angle φ. Half of the length of the edge is subtracted from this distance, and what remains is the Euclidean distance to the edge.

Using the same initialization scheme as above, the algorithm also manages the rotated rectangles in figure 5.15. Nevertheless, the general problems mentioned in the chapter on shell clustering algorithms also apply to rectangular shapes, i.e. a large number of clusters or barely separated clusters lead to undesired results like in figure 5.16.

Although this algorithm yields quite good results in comparison to the previously mentioned algorithms for rectangles, the drawback is the necessity to compute the prototypes on a numerical basis.

An exact description of rectangular shapes is not only possible with the maximum norm, but also with the p-norm with $p = 1$. In contrast to the maximum norm, the *unit circle* with respect to the p-norm with $p = 1$ is not an axes-parallel square, but a square rotated by 45 degrees, standing on one of its corners. The necessity of a numerical computation of the prototypes remains, the principal problems caused by the absolute value cannot be overcome. Instead of a direct computation of the prototypes for this special distance function other strategies to optimize the objective function like genetic algorithms can also be considered, however without great success as reported in [13]. A direct solution for determining the prototypes is usually faster and yields better results. The large number of local minima of the objective functions in shell clustering seems to cause problems for genetic algorithms.

5.2   The fuzzy c-rectangular shells algorithm

The fuzzy c-rectangular shells algorithm uses a distance measure, that reflects the Euclidean distance to the nearest rectangles edge. The contour of a rectangle can be assembled by four lines, each described by a normal equation $(x - p)^\top n = 0$, in which $p \in \mathbb{R}^2$ is a point of the considered line and $n \in \mathbb{R}^2$ the normal vector of the line. If n is a unit normal vector, the expression $|(x - p)^\top n|$ yields the Euclidean distance of a point x to the line. Initially, we have to formalize the four boundary lines of a rectangle.

As shown in figure 5.17, we denote the centre of the rectangle by $v \in \mathbb{R}^2$ and the edge lengths by $2r_0$ and $2r_1$, $r = (r_0, r_1) \in \mathbb{R}^2$. Furthermore, φ is the angle between the positive x-axis and the positive r_0-axis of the rectangle. We enumerate the lines (that will become the edges of the rectangle) counter-clockwise, beginning with zero at the right line ($\varphi = 0$ assumed). The points and normal vectors of the lines are numbered in the same way. For the unit normal vectors n_s, $s \in \mathbb{N}^*_{<4}$, we require $(v - p_s)^\top n_s > 0$. In this way, all normal vectors are *directed* towards the centre of the rectangle. This means:

$$n_s = \begin{pmatrix} -\cos(\varphi + \frac{s\pi}{2}) \\ -\sin(\varphi + \frac{s\pi}{2}) \end{pmatrix}, \text{ or}$$

$$n_0 = \begin{pmatrix} -\cos(\varphi) \\ -\sin(\varphi) \end{pmatrix} = -n_2, \; n_1 = \begin{pmatrix} \sin(\varphi) \\ -\cos(\varphi) \end{pmatrix} = -n_3.$$

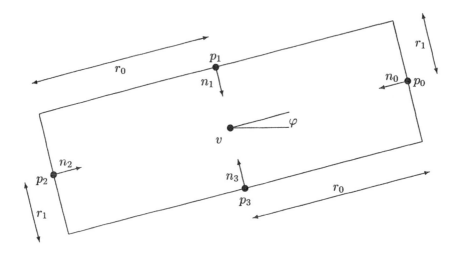

Figure 5.17: FCRS parameters

The points p_s can be placed anywhere on the corresponding lines. But after fixing the normal vectors, the obvious way to define p_s would be

$$p_s = v - r_{s \bmod 2} n_s \quad \text{for all } s \in \mathbb{N}^*_{<4}.$$

Since the normal vectors point towards the centre v, $(x - p_s)^\top n_s$ yields a positive value only if the vector x lies on the side of the line near the rectangle centre. We will call this value the *directed* distance. Since n_s is normalized, we can rewrite the directed distance as

$$\begin{aligned}(x - p_s)^\top n_s &= (x - (v - r_{s \bmod 2} n_i))^\top n_s = (x - v)^\top n_s + r_{s \bmod 2} n_s^\top n_s \\ &= (x - v)^\top n_s + r_{s \bmod 2}\end{aligned}$$

for $s \in \mathbb{N}^*_{<4}$. Evaluating the minimum of all directed distances, we get a positive/zero/negative value, if x lies inside/on/outside the rectangle. The graph of the minimum looks like a pyramid, shown by figure 5.18. The absolute value of the minimum is the Euclidean distance between x and the nearest edge of the rectangle (figure 5.19). Therefore, we define the Fuzzy c-Rectangular Shell distance measure as

$$\begin{aligned}d^2 : \mathbb{R}^2 \times C &\to \mathbb{R}, \\ (x, (v, r, \varphi)) &\mapsto \left(\min\{(x - v)^\top n_s + r_{s \bmod 2} \mid s \in \mathbb{N}^*_{<4}\}\right)^2, \quad (5.4)\end{aligned}$$

where $C := \mathbb{R}^2 \times \mathbb{R}^2 \times \mathbb{R}$. Taking the absolute value explicitly is not necessary, as the squared distance measure is used by the algorithm.

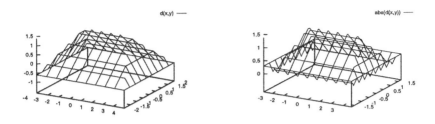

Figure 5.18:   Figure 5.19:

Note that the lines are clipped (to true edges) by the min-function, so they have no infinite expansion as in the FCV algorithm. But the use of the minimum function prevents the prototype parameters from being determined explicitly. We cannot solve for the parameters that are used inside the minimum function. The schematic procedure (differentiate and solve for the parameters) that was used for all the other clustering algorithms fails if we want to calculate the parameters directly and avoid iterative Newton-like methods. What can we do to keep the additional computational costs small? A human probably would simplify the problem by dividing the data into four clusters: one cluster for each edge. The optimization of four lines seems to be easier than the ad hoc calculation of the whole rectangle. At first glance, this is the same problem as it can be solved by line detection algorithms. But there are some differences:

- Two membership degrees are assigned to each data vector: one for the rectangle and one for the particular edges of every rectangle. (This could be realized by using a modified distance measure for the line detection algorithm, that takes into account the membership degree to the rectangle as a scaling factor.)

- There are several relationships between the edges of a rectangle. Two edges must be parallel, meeting the other pair at a right angle. The optimization has to reflect these constraints.

- To simplify the problem effectively, a human would prefer a clear partition of the data vectors into four clusters for each rectangle. Uncertainties near the vertices are quite unproblematic. Those data vectors lie on both adjacent lines.

All these observations are justified by the right-angled shape of a rectangle and suggest the use of a second – possibly hard – partition for the edges. There is a fundamental difference between other (fuzzy) shell clustering algorithms and rectangular shell clustering. In the case of rectangles it seems to be insufficient to identify only the cluster (rectangle) to which a datum should be assigned with a certain membership degree. In a second step one has to associate one of the four edges of the identified rectangle with the datum. The first (fuzzy) partition can be obtained in the same way as it is done in all the other algorithms. The second (possibly hard) partition is produced by the minimum function itself: we associate the data vector with the line that is responsible for the minimum value. By generating the second partition in this way, we actually rewrite the minimum function in another form. Using the Kronecker symbol δ ($\delta_{i,j} = 1$ if $i = j$, $\delta_{i,j} = 0$ otherwise) we define for every $s \in \mathbb{N}^*_{<4}$:

$$\min_s^h : \mathbb{R}^4 \quad \to \quad \{0,1\}, \quad (a_0, a_1, a_2, a_3) \mapsto \delta_{s,m} \tag{5.5}$$
$$\text{with } a_m = \min\{a_0, a_1, a_2, a_3\}.$$

For example $\min_2^h(7,8,4,6) = 1$ and $\min_s^h(7,8,4,6) = 0$ for $s \in \{0,1,3\}$, because the minimum of a_0, a_1, a_2 and a_3 is a_2. We must bear in mind, however, that only one of the four minimum functions has to return the value 1, even if multiple a_i are minimal. In this case we have to choose (randomly) one $s \in \{j \in \mathbb{N}^*_{<4} \mid a_j = \min\{a_0, a_1, a_2, a_3\}\}$. For example, $\min_0^h(4,4,8,9) = 1 = \min_1^h(4,4,8,9)$ is not allowed. The kind of minimum functions we have in mind must satisfy the equation $1 = \sum_{s=0}^3 \min_s^h(a_0, a_1, a_2, a_3)$. In this way, $\min\{a_0, a_1, a_2, a_3\} = \sum_{i=0}^3 a_i \min_i^h(a_0, a_1, a_2, a_3)$ holds. We interpret $\min_s^h(a_0, a_1, a_2, a_3)$ as the *degree of minimality* of a_s referring to a_0, a_1, a_2 and a_3. Let $\Delta_s(x, k)$ be the (directed) distance of a point x to the prototype k and

$$\tilde{u}_{i,j,s} := \min_s^h((\Delta_t(x_j, k_i))_{t \in \mathbb{N}^*_{<4}}). \tag{5.6}$$

Then $\tilde{u}_{i,j,s}$ denotes the membership degree of data vector x_j to edge s of cluster k_i. And $\tilde{u}_{i,j \in \mathbb{N}_{\leq n}, s \in \mathbb{N}^*_{<4}}$ is a partition that assigns data vectors to rectangle edges. With these terms we rewrite the distance (5.4) between a data vector x_j and a rectangle prototype $k_i = (v, r, \varphi)$ as:

$$d_{i,j}^2 = \left(\sum_{s=0}^3 \tilde{u}_{i,j,s}(x_j - v)^\top n_s + r_{s \bmod 2}\right)^2$$

$$= \sum_{s=0}^3 \tilde{u}_{i,j,s}\left((x_j - v)^\top n_s + r_{s \bmod 2}\right)^2. \tag{5.7}$$

(This follows from the fact, that products of $\tilde{u}_{i,j,s}$ with different s evaluate to zero and since $\tilde{u}_{i,j,s}^2 = \tilde{u}_{i,j,s} \in \{0,1\}$.)

Theorem 5.5 (FCRS prototypes)
Let $D := \mathbb{R}^2$, $X = \{x_1, x_2, \ldots, x_n\} \subseteq D$, $C := \mathbb{R}^2 \times \mathbb{R}^2 \times \mathbb{R}$ be the set of possible (parameterized) rectangles (as defined previously), $m \in \mathbb{R}_{>1}$,

$$\Delta_s \quad : \quad D \times C \to \mathbb{R}_{\geq 0}, \quad (x, (v, r, \varphi)) \mapsto \left((x - v)^\top n_s + r_{s \bmod 2}\right),$$

for all $s \in \mathbb{N}^*_{<4}$ and

$$d^2 \quad : \quad D \times C \to \mathbb{R}_{\geq 0}, \quad (x, k) \mapsto \sum_{s=0}^{3} \min_s \left((\Delta_t(x, k))_{t \in \mathbb{N}^*_{<4}}\right) \Delta_s^2(x, k),$$

with $n_s = (-\cos(\varphi + \frac{s\pi}{2}), -\sin(\varphi + \frac{s\pi}{2}))^\top$
 If J is minimized by $f : X \to F(K)$ with respect to all probabilistic cluster partitions $X \to F(K)$ with $K = \{k_1, k_2, \ldots, k_c\} \in R$ and given memberships $f(x_j)(k_i) = u_{i,j}$, so with $k_i = (v, r, \varphi)$, $v = \mu_0 n_0 + \mu_1 n_1$, $\tilde{u}_{i,j,s} = \min_s^h \left((\Delta_t(x_j, k_i))_{t \in \mathbb{N}^*_{<4}}\right)$ and $r = (r_0, r_1)$:

$$\mu_s \quad = \quad S_{i,s+2}\, n_{s+2} + S_{i,s}\, n_s \tag{5.8}$$

$$r_s \quad = \quad S_{i,s+2}\, n_{s+2} - S_{i,s}\, n_s \tag{5.9}$$

$$\varphi \quad = \quad \text{atan2}(n_0) \tag{5.10}$$

n_0 *is   the normalized eigenvector of A_i associated with the smallest eigenvalue* (5.11)

$$A_i \quad = \quad \sum_{j=1}^{n} u_{i,j}^m \left(a_{i,j,0} a_{i,j,0}^\top + \tau(a_{i,j,1}) \tau(a_{i,j,1})^\top\right)$$

$$S_{i,s} \quad = \quad \frac{\sum_{j=1}^{n} u_{i,j}^m\, \tilde{u}_{i,j,s}\, x_j^\top}{2 \sum_{j=1}^{n} u_{i,j}^m\, \tilde{u}_{i,j,s}}$$

$$a_{i,j,s} \quad = \quad (\tilde{u}_{i,j,s} - \tilde{u}_{i,j,s+2}) x_j - 2\tilde{u}_{i,j,s} S_{i,s} + 2\tilde{u}_{i,j,s+2} S_{i,s+2}$$

$$\tau \quad : \quad \mathbb{R}^2 \to \mathbb{R}^2, \quad (x, y) \mapsto (-y, x)$$

for all $s \in \{0, 1\}$.

Proof: For the objective function $J := \sum_{j=1}^{n} \sum_{i=1}^{c} u_{i,j}^m d^2(x_j, k_i)$ to have a minimum it is necessary that the partial derivatives are zero. The use of the hard minimum functions $\min_s^h$ allows us to simplify the distance function as in (5.5). With $k_i = (v, r, \varphi)$ and $n_s = -n_{s+2}$ the differentiation of J with respect to the (half) edge lengths r_s leads for $s \in \{0, 1\}$ to:

$$\frac{\partial J}{\partial r_s} = 0 \quad \Rightarrow \quad \sum_{j=1}^{n} u_{i,j}^m \left(\tilde{u}_{i,j,s}((x_j - v)^\top n_s + r_s) + \right.$$

$$\left. \tilde{u}_{i,j,s+2}((x_j - v)^\top n_{s+2} + r_s) \right) = 0$$

$$\Leftrightarrow \quad \sum_{j=1}^{n} u_{i,j}^m \left((\tilde{u}_{i,j,s} - \tilde{u}_{i,j,s+2})(x_j - v)^\top n_s + \right.$$

$$\left. (\tilde{u}_{i,j,s} + \tilde{u}_{i,j,s+2}) r_s \right) = 0. \tag{5.12}$$

For the derivative of the centre v we get for all $\xi \in \mathbb{R}^2$ by applying $n_s = -n_{s+2}$ twice:

$$\frac{\partial J}{\partial v} = 0$$

$$\Rightarrow \quad \sum_{j=1}^{n} u_{i,j}^m \sum_{s=0}^{3} \tilde{u}_{i,j,s}((x_j - v)^\top n_s + r_{s \bmod 2}) \xi^\top n_s = 0$$

$$\Leftrightarrow \quad \sum_{j=1}^{n} u_{i,j}^m \sum_{s=0}^{1} \left((\tilde{u}_{i,j,s} + \tilde{u}_{i,j,s+2})(x_j - v)^\top n_s + \right.$$

$$\left. (\tilde{u}_{i,j,s} - \tilde{u}_{i,j,s+2}) r_s \right) \xi^\top n_s = 0.$$

Especially for $\xi = n_0$ and $\xi = n_1$ one of the terms becomes zero, because of $n_s^\top n_s = 1$ and $n_{1-s}^\top n_s = 0$ for $s \in \{0,1\}$. This leads to

$$\sum_{j=1}^{n} u_{i,j}^m \left((\tilde{u}_{i,j,s} + \tilde{u}_{i,j,s+2})(x_j - v)^\top n_s + \right.$$

$$\left. (\tilde{u}_{i,j,s} - \tilde{u}_{i,j,s+2}) r_s \right) = 0 \tag{5.13}$$

for $s \in \{0,1\}$. Choose $s \in \{0,1\}$. The resulting equations (5.12) and (5.13) only differ in the sign of some $\tilde{u}_{i,j,s}$ terms. From the addition and subtraction of (5.12) and (5.13)

$$\sum_{j=1}^{n} u_{i,j}^m \left(\tilde{u}_{i,j,s}(x_j - v)^\top n_s + \tilde{u}_{i,j,s} r_s \right) = 0,$$

$$\sum_{j=1}^{n} u_{i,j}^m \left(\tilde{u}_{i,j,s+2}(x_j - v)^\top n_s - \tilde{u}_{i,j,s+2} r_s \right) = 0$$

can be obtained. Another simplification can be achieved by denoting v as a linear combination of n_0 and n_1, for example $v = \mu_0 n_0 + \mu_1 n_1$. Then the product $v^\top n_s$ equals μ_s. By rearranging both equations for μ_s, we get:

$$\mu_s = \frac{\sum_{j=1}^{n} u_{i,j}^m \tilde{u}_{i,j,s} x_j^\top n_s}{\sum_{j=1}^{n} u_{i,j}^m \tilde{u}_{i,j,s}} + r_s,$$

$$\mu_s = \frac{\sum_{j=1}^{n} u_{i,j}^m \tilde{u}_{i,j,s+2} x_j^\top n_s}{\sum_{j=1}^{n} u_{i,j}^m \tilde{u}_{i,j,s+2}} - r_s.$$

Using our notation in the theorem we obtain $\mu_s = S_{i,s} n_s + S_{i,s+2} n_s$ by adding these two equations, yielding (5.8) as the result. By subtraction we get (5.9) analogously. Now we can replace r_s and v in our objective function, where only the angle φ remains unknown:

$$\sum_{j=1}^{n} u_{i,j}^m \sum_{s=0}^{3} \tilde{u}_{i,j,s} ((x_j - v)^\top n_s + r_{s \bmod 2})^2$$

$$= \sum_{j=1}^{n} u_{i,j}^m \sum_{s=0}^{1} \left((\tilde{u}_{i,j,s} - \tilde{u}_{i,j,s+2})(x_j^\top n_s - \mu_s) \right.$$

$$\left. + (\tilde{u}_{i,j,s} + \tilde{u}_{i,j,s+2}) r_s \right)^2$$

$$= \sum_{j=1}^{n} u_{i,j}^m \sum_{s=0}^{1} \left((\tilde{u}_{i,j,s} - \tilde{u}_{i,j,s+2})(x_j - S_{i,s+2} - S_{i,s}) n_s \right.$$

$$\left. + (\tilde{u}_{i,j,s} + \tilde{u}_{i,j,s+2})(S_{i,s+2} - S_{i,s}) n_s \right)^2$$

$$= \sum_{j=1}^{n} u_{i,j}^m \sum_{s=0}^{1} \left(((\tilde{u}_{i,j,s} - \tilde{u}_{i,j,s+2}) x_j \right.$$

$$\left. - 2\tilde{u}_{i,j,s} S_{i,s} + 2\tilde{u}_{i,j,s+2} S_{i,s+2}) n_s \right)^2$$

$$= \sum_{j=1}^{n} u_{i,j}^m \sum_{s=0}^{1} (n_s^\top a_{i,j,s})(a_{i,j,s}^\top n_s).$$

For further simplification we use the following relationship between n_0 and n_1: $\forall x \in \mathbb{R}^2 : x^\top n_1 = \tau(x)^\top n_0$. We continue:

$$= \sum_{j=1}^{n} u_{i,j}^{m} \left((n_0^{\top} a_{i,j,0})(a_{i,j,0}^{\top} n_0) \right.$$

$$\left. + (n_0^{\top} \tau(a_{i,j,1}))(\tau(a_{i,j,1})^{\top} n_0) \right)$$

$$= n_0^{\top} \left(\sum_{j=1}^{n} u_{i,j}^{m} \left(a_{i,j,0} a_{i,j,0}^{\top} + \tau(a_{i,j,1}) \tau(a_{i,j,1})^{\top} \right) \right) n_0$$

$$= n_0^{\top} A_i n_0.$$

But this expression is minimal when n_0 is the eigenvector of matrix A_i with the smallest eigenvalue. From n_0 we obtain φ by applying the atan-function (5.10). ∎

Unfortunately the use of the crisp degrees of minimality leads to a convergence problem. Figure 5.20 shows the top and left edge of a rectangle. On the left hand side of the diagonal, all data vectors are assigned to the left edge, on the right hand side to the top edge. In this example a data vector lies exactly on this diagonal, that means it has the same distance to both of the edges. Because of the hard degrees of minimality, we have to decide to which edge this vector should be assigned. If we choose the top line, figure 5.21 shows a possible parameter configuration for the next iteration step. Now the data vector lies on the left-hand side of the diagonal and in the next step it will be assigned to the left edge, leading to figure 5.22. It is evident that this vector causes two alternating states. That prevents the algorithm from converging. (Depending on the data set this problem actually occurs in practice.) While the fuzzy partition matrix U is still subject to greater changes in the first iteration steps, the described problem does not have strong effects. The more the algorithm reaches its final result, the more the probability for such an alternation increases.

As the introduction of the fuzzy partition matrix helped the hard c-means algorithm to avoid local minima, fuzzy minimum functions should avoid the convergence problem of the fuzzy c-rectangular shells algorithm. We substitute the hard minimum functions by fuzzy ones $\min_i : \mathbb{R}^4 \to [0,1]$, which still satisfy $1 = \sum_{s=0}^{3} \min_s(a_0, a_1, a_2, a_3)$ for all $a \in \mathbb{R}^4$. We need this constraint, because every data vector has to keep the same weight in our objective function. (Otherwise the objective function could be minimized by setting all degrees of minimality to zero.) There are many possible fuzzifications, all influencing the result of FCRS in a different way. If we compare the fuzzy partition matrix U and the degrees of minimality we recognize similarities, suggesting to us a first fuzzification: Assuming a_0

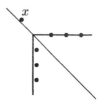

Figure 5.20: Data vector x lies on the diagonal

Figure 5.21: Data vector x is assigned to the top edge

Figure 5.22: Data vector x is assigned to the left edge

is the minimum of a_0, a_1, a_2 and a_3, we can interpret $d_i := a_i - a_0$ as the distance of a_i from the minimum a_0, $i \in \mathbb{N}^*_{<4}$. Adding a positive $\varepsilon \in \mathbb{R}$ to every distance d_i, we can calculate degrees of minimality by $\min_s = \left(\sum_{i=0}^{3} \frac{(d_s + \varepsilon)^2}{(d_i + \varepsilon)^2} \right)^{-1}$, which is an analogous calculation as the one used for the partition matrix U. We define

$$\min_s : \mathbb{R}^p \to [0, 1], \quad a \mapsto \frac{1}{\sum_{i=0}^{p-1} \frac{(a_s - m + \varepsilon)^2}{(a_i - m + \varepsilon)^2}} \tag{5.14}$$
$$\text{where} \quad m = \min\{a_j | j \in \mathbb{N}^*_{<p}\}$$

for $s \in \mathbb{N}^*_{<4}$. If ε approaches zero, the fuzzy minimum functions converge towards the hard functions, if ε approaches infinity, the fuzzy minimum functions converge towards $\frac{1}{n}$, where n is the dimension of the domain of the fuzzy minimum function. Figure 5.23 shows $f(a_0, a_1, \varepsilon) = \min_0(a_0, a_1)$ for different values of ε.

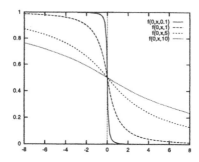

Figure 5.23: Influence of the parameter ε on the fuzzification

By choosing the minimum functions as too fuzzy, we take into account too many data vectors of the neighbouring edges for estimating the actual edge and the estimated edges lie inside (not on) the rectangle. On the other hand, if the range in which the degrees of minimality float from one edge to another is too small, we cannot avoid an alternating sequence as shown in figures 5.20 to 5.22. We have to choose ε depending on the density of the data vectors. Denoting the average distance between two neighbouring data vectors by δ, a fuzzification involving an area of δ to 2δ should work fine for most cases. (At least it does in our examples.)

In computer aided image recognition the data vectors often lie on a fixed grid. Then δ need not be evaluated by a complex expression like

$$\delta \approx \frac{\sum_{i=1}^{n} \min\{\, \|x_i - x_j\| \mid j \in \mathbb{N}_{\leq n}\setminus\{i\}\,\}}{n} \tag{5.15}$$

but is given by the image resolution.

In general, we assume that the rectangle's edge lengths are large compared to the distance between neighbouring data vectors. If x is a data vector lying on one of the rectangle's edges, with a distance d to the nearest adjacent edge, then we approximate the degree of membership u of x belonging to the corresponding edge by

$$u \approx \frac{1}{\frac{\varepsilon^2}{\varepsilon^2} + \frac{\varepsilon^2}{(d+\varepsilon)^2}} = \frac{1}{1 + \frac{\varepsilon^2}{(d+\varepsilon)^2}} \quad \Rightarrow \quad \varepsilon \approx \frac{\sqrt{u(1-u)} + (1-u)}{2u - 1}\, d\,.$$

In this approximation we ignored two distances because they were small compared to the remaining terms. Furthermore, only $u > \frac{1}{2}$ makes sense, as u denotes the membership degree to the edge on which x is located.

If we want a clear classification of the data vector at a distance of $d = 2\delta$ with $u = 0.95$, then equation (5.23) suggests $\varepsilon \approx \frac{3\delta}{5}$. In this way, we can choose, depending on the image resolution, an adequate fuzzification of our minimum functions.

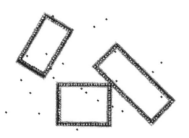

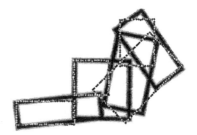

Figure 5.24: FCRS partition (P)   Figure 5.25: FCRS partition (P)

For the following examples, ten iterations of FCM followed by ten iterations of FCSS provided the FCRS initialization. (We use FCSS instead of FCS, AFCS or FCES only because of computational costs. FCQS supports some cluster shapes such as hyperbolas that are difficult to convert into rectangular shapes.) All algorithms used a fuzzifier value of 2. Half of the radii determined by the FCSS algorithm were used for the half edge lengths of the FCRS algorithm, because the detected circles were almost always too large. Like all other shell clustering algorithms FCRS is highly dependent on the initialization. The detection of longitudinal rectangles for example cannot be supported by the FCSS algorithm, instead of the FCSS iterations the number of FCM iterations could be increased or the GK algorithm could be used.

Figures 5.24 and 5.25 show data sets with rotated and overlapping rectangles of different sizes. Although there is some noise in figure 5.24 the rectangles have been recognized, which can be improved by a possibilistic run. Excessive noise in the data sets causes a lot of local minima, and similar to most of the shell clustering algorithms possibilistic FCRS then fails to detect the image structure. Figure 5.25 seems to present the FCRS algorithm with another problem. Following the stepwise development of this partition it gives the impression that the data vectors inside the rectangle should have higher influence on the resulting cluster: It looks like all clusters try to reduce the distance to the data vectors outside the cluster, but intuitively they should better take the data vectors inside into account. However, this is not aided by the fuzzy minimum functions, which we can see with a closer look at the lower left cluster in figure 5.25. The horizontal extent of this cluster is too large, the right edge is approximated by another cluster.

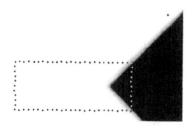

Figure 5.26: Membership to right edge, using fuzzy minimum function (5.14)

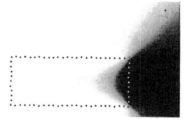

Figure 5.27: Membership to right edge, using fuzzy minimum function (5.16)

Figure 5.26 shows a similar rectangle and the membership degree to the right edge. In a distance roughly half the size of the length of the smaller edge, the high membership degrees terminate abruptly. In figure 5.25 the data vectors lie outside the range of influence of the right edge. A final modification of the minimum functions should amend this problem:

$$\min_s : \mathbb{R}^p \to [0,1], \quad a \mapsto \frac{1}{\sum_{i=0}^{p-1} \frac{(|m|+\exp(a_s-m)-1+\varepsilon)^2}{(|m|+\exp(a_i-m)-1+\varepsilon)^2}} \tag{5.16}$$
$$\text{where } m = \min\{a_j | j \in \mathbb{N}^*_{<p}\}.$$

Compared to variation (5.14), we translate the distances by the absolute value of the minimum and reweight the distance using the exponential function. If the minimum is near zero, this modification is similar to (5.14). Therefore we choose ε in the same way as before. These changes lead to fuzzier minimality degrees for non-zero minimum values, i.e. for data vectors far away from the contour.

Figure 5.27 shows the degrees of minimality achieved by the new functions. Along the edges of the rectangle there are still high memberships, whereas they are much more fuzzy in the distance. Even inside the rectangle there is a smoother crossing and the sphere of influence has grown. This is useful in our case, because we do not want to clearly assign data vectors inside the rectangle to the edges while we do not know its final orientation. Using this fuzzification, the FCRS algorithm detects all rectangles in figure 5.25 correctly (a good initialization presumed). For the following examples, we will keep this last modification of the fuzzy minimum functions.

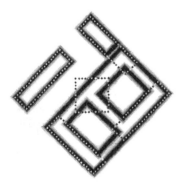

Figure 5.28: FCRS partition (P)

Figure 5.29: FCRS partition (P)

The danger of converging in a local minimum with an algorithm to detect rectangles is greater than with an algorithm to detect spheres or ellipses. A circle can be defined unambiguously by three points. But if a depicted rectangle consists only of some of its edges, there are many possible rectangles that approximate the remaining data vectors. Furthermore, if some edges are parallel they could easily be exchanged between different rectangle-clusters, which leads to a strong local minimum. This situation can often be observed, if the angles of rotation differ about 90 degrees.

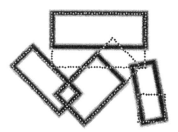 

Figure 5.30: FCRS partition (P)   Figure 5.31: FCRS partition (P)

Figure 5.28 shows an example. Although the data vectors are approximated quite well, the rectangle parameters are not correct. In this case, the FCSS iteration steps are responsible for the wrong partition. The detected circles assign data vectors that are far away from each other to the same clusters. If we leave out the FCSS initialization, the FCRS algorithm detects the clusters much better. Despite the fact that the dataset in figure 5.29 looks much more complicated, the FCRS algorithm discovers all rectangles after ten FCM and ten FCS iterations. The fact that edges of different rectangles never lie parallel to each other makes the detection easier. We want to bear in mind that strong local minima with parallel edges are not a particular problem for the FCRS algorithm, but are for any algorithm detecting rectangular shapes. (A conventional line detection algorithm would lump together several edges in figure 5.28. Since there are many collinear edges, there are fewer lines than edges. The superfluous lines would be fixed randomly by some (noisy) data vectors. We had to ignore some detected lines and to split others to find the correct partition.)

Because of the Euclidean distance provided by the FCRS distance function, the results can sometimes be improved by increasing the fuzzifier. Although FCRS does not detect the rectangles in figure 5.30 with a fuzzifier $m = 2.0$ because of the parallel edges, it does detect them with an increased fuzzifier $m = 4.0$ in figure 5.31. Of course, the Euclidean distance measure cannot guarantee an improvement of the clustering result at a higher fuzzifier, especially if the objective function contains strong local minima, like in figure 5.28.

For implementational remarks on FCRS, see page 153.

5.3 The fuzzy c-2-rectangular shells algorithm

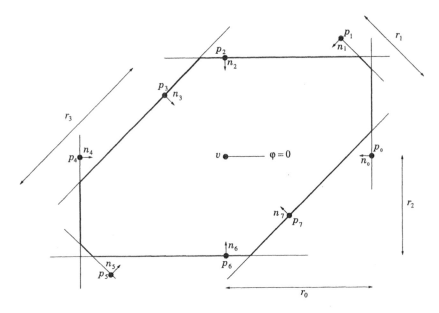

Figure 5.32: Definition of the FC2RS cluster

In some applications the ability to detect more complex shapes might be useful. Taking a closer look at the parameter set of FCRS, the concept can be easily generalized to polygons other than rectangles. In the derivation of the FCRS algorithm, we made use of the fact that the edge's normal vectors vary in steps of 90 degrees from edge to edge. If we decrease this angle we could realize more complex shapes. Actually, every angle is possible, as long as it is a divisor of 360 degrees. In a special case an algorithm to detect triangular shapes might be useful, for which we would use an angle of 120 degrees. But we want to keep the ability of detecting rectangles, therefore, we must also reach 90, 180 and 270 degrees. The greatest angle less than 90 degrees satisfying these conditions is 45 degrees.

By using steps of 45 degrees we get eight edges realizing the cluster shape. Again, we denote the cluster centre by v, the angle of rotation by φ and now the four half edge lengths by r_0 to r_3, as displayed in figure 5.32. (Now, a cluster is completely described by a triple $(v, r, \varphi) \in \mathbb{R}^2 \times \mathbb{R}^4 \times \mathbb{R}$.) The boundary lines are defined by

$$(x - p_s)^\top n_s = 0 \text{ where } n_s = \begin{pmatrix} -\cos(\varphi + \frac{s\pi}{4}) \\ -\sin(\varphi + \frac{s\pi}{4}) \end{pmatrix} \text{ and } p_s = v - r_{s \bmod 4} \, n_s$$

for $s \in \mathbb{N}^*_{<8}$. We can illustrate the possible cluster shapes by overlaying two rectangles with aligned centres. The shortest closed path along the rectangle's edges describes the contour of the cluster. In particular, if one of the rectangles is very large the resulting form matches the shape of the small rectangle. Figure 5.33 shows some examples for detectable FC2RS shapes, marked by the thick lines. The complex shapes are appropriate for circles or ellipses, therefore FC2RS might handle complicated scenes with rectangles and ellipses correctly. (The name of the algorithm *FC2RS* originates from the visualization of the cluster's shape with the help of two rectangles.)

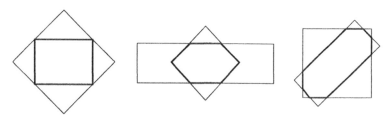

Figure 5.33: Cluster shapes of the FC2RS algorithm

The distance function evaluates the minimum of the distance to all edges and yields its absolute value. To find an explicit expression for the prototypes we formalize the minimum function as in (5.5).

Theorem 5.6 (FC2RS prototypes)
Let $D := \mathbb{R}^2$, $X = \{x_1, x_2, \ldots, x_n\} \subseteq D$, $C := \mathbb{R}^2 \times \mathbb{R}^4 \times \mathbb{R}$, $c \in \mathbb{N}$, $R := \mathcal{P}_c(C)$, J corresponding to (1.7) with $m \in \mathbb{R}_{\geq 1}$ and

$$\Delta_s \quad : \quad D \times C \to \mathbb{R}_{\geq 0}, \quad (x, (v, r, \varphi)) \mapsto ((x - v)^\top n_s + r_{s \bmod 4}),$$

$$d^2 \quad : \quad D \times C \to \mathbb{R}_{\geq 0}, \quad (x, k) \mapsto \sum_{s=0}^{7} \min_s((\Delta_t(x, k))_{t \in \mathbb{N}^*_{<8}}) \Delta_s^2(x, k),$$

and $n_s = (-\cos(\varphi + \frac{s\pi}{4}), -\sin(\varphi + \frac{s\pi}{4}))^\top$ for all $s \in \mathbb{N}^*_{<8}$. If J is minimized by $f : X \to F(K)$ w.r.t. all probabilistic cluster partitions

$X \rightarrow F(K)$ *with* $K = \{k_1, k_2, \ldots, k_c\} \in R$ *and given memberships* $f(x_j)(k_i) = u_{i,j}$, *so with* $k_i = (v, r, \varphi)$, $v = \lambda n_0 + \mu n_2$, $r = (r_0, r_1, r_2, r_3)$, $\tilde{u}_{i,j,s} = \min_s^h ((\Delta_t(x_j, k_i))_{t \in \mathbb{N}_{<8}^*})$:

$$
\begin{aligned}
\varphi &= \operatorname{atan2}(n_0) \\
\lambda &= \tilde{l}^\top n_0 \\
\mu &= \tilde{m}^\top n_0 \\
r_s &= \tilde{r}_s^\top n_0
\end{aligned}
$$

where

$$
\begin{aligned}
\tilde{l} &= \frac{1}{q}((\sqrt{2}R_1 a_{i,1} + a_{i,0} + R_2 a_{i,2})(\alpha_{j,3} + \alpha_{i,2}) - \\
&\quad (\sqrt{2}R_3 a_{i,3} - a_{i,0} + R_2 a_{i,2})(\alpha_{i,1} + \alpha_{i,2})) \\
\tilde{m} &= \frac{1}{q}((\sqrt{2}R_1 a_{i,1} + a_{i,0} + R_2 a_{i,2})(\alpha_{i,3} + \alpha_{i,0}) - \\
&\quad (\sqrt{2}R_3 a_{i,3} - a_{i,0} + R_2 a_{i,2})(\alpha_{i,1} + \alpha_{i,0})) \\
q &= (\alpha_{i,1} + \alpha_{i,2})(\alpha_{i,3} + \alpha_{i,0}) + (\alpha_{i,3} + \alpha_{i,2})(\alpha_{i,1} + \alpha_{i,0}) \\
\tilde{l}_0 &= -\tilde{l}_4 = \tilde{l}, \; \tilde{l}_1 = -\tilde{l}_5 = \frac{\tilde{l} + \tilde{m}}{\sqrt{2}}, \; \tilde{l}_2 = -\tilde{l}_6 = \tilde{m}, \; \tilde{l}_3 = -\tilde{l}_7 = \frac{-\tilde{l} + \tilde{m}}{\sqrt{2}}
\end{aligned}
$$

n_0 *is the normalized eigenvector of A associated with the smallest eigenvalue*

$$
A = \sum_{j=1}^n u_{i,j}^m \sum_{s=0}^7 \tilde{u}_{i,j,s}(R_s x_j - \tilde{l}_s + \tilde{r}_{s \bmod 4})(R_s x_j - \tilde{l}_s + \tilde{r}_{s \bmod 4})^\top
$$

and for all $s \in \mathbb{N}_{<4}^*$

$$
\tilde{r}_s = -\frac{R_s \sum_{j=1}^n u_{i,j}^m (\tilde{u}_{i,j,s} - \tilde{u}_{i,j,s+4}) x_j}{\sum_{j=1}^n u_{i,j}^m (\tilde{u}_{i,j,s} + \tilde{u}_{i,j,s+4})} + \frac{\sum_{j=1}^n u_{i,j}^m (\tilde{u}_{i,j,s} - \tilde{u}_{i,j,s+4})}{\sum_{j=1}^n u_{i,j}^m (\tilde{u}_{i,j,s} + \tilde{u}_{i,j,s+4})} \tilde{l}_s
$$

$$
a_{i,s} = \sum_{j=1}^n u_{i,j}^m (\tilde{u}_{i,j,s} + \tilde{u}_{i,j,s+4}) x_j - \\
\frac{\sum_{j=1}^n u_{i,j}^m (\tilde{u}_{i,j,s} - \tilde{u}_{i,j,s+4})}{\sum_{j=1}^n u_{i,j}^m (\tilde{u}_{i,j,s} + \tilde{u}_{i,j,s+4})} \sum_{j=1}^n u_{i,j}^m (\tilde{u}_{i,j,s} - \tilde{u}_{i,j,s+4}) x_j
$$

$$
\alpha_{i,s} = \sum_{j=1}^n u_{i,j}^m (\tilde{u}_{i,j,s} + \tilde{u}_{i,j,s+4}) - \frac{\left(\sum_{j=1}^n u_{i,j}^m (\tilde{u}_{i,j,s} - \tilde{u}_{i,j,s+4})\right)^2}{\sum_{j=1}^n u_{i,j}^m (\tilde{u}_{i,j,s} + \tilde{u}_{i,j,s+4})}
$$

and for all $s \in \mathbb{N}_{<8}^*$

$$R_s = \begin{pmatrix} \cos(\frac{s\pi}{4}) & \sin(\frac{s\pi}{4}) \\ -\sin(\frac{s\pi}{4}) & \cos(\frac{s\pi}{4}) \end{pmatrix}$$

holds.

Proof: For the objective function $J := \sum_{j=1}^n \sum_{i=1}^c u_{i,j}^m d^2(x_j, k_i)$ to have a minimum it is necessary that the partial derivatives vanish. The use of the hard minimum functions (5.5) allows us to simplify the distance function in the same way as in (5.7). With $k_i = (v, r, \varphi)$ and $n_s = -n_{s+4}$ we get in analogy to theorem 5.5, for all $s \in \mathbb{N}_{<4}^*$ and $\xi \in \mathbb{R}^2$:

$$\frac{\partial J}{\partial r_s} = 0 \quad \Rightarrow$$

$$\sum_{j=1}^n u_{i,j}^m \left((\tilde{u}_{i,j,s} - \tilde{u}_{i,j,s+4})(x_j - v)^\top n_s + (\tilde{u}_{i,j,s} + \tilde{u}_{i,j,s+4})r_s \right) = 0 \,,$$

$$\frac{\partial J}{\partial v} = 0 \quad \Rightarrow$$

$$\sum_{j=1}^n u_{i,j}^m \sum_{s=0}^3 \left((\tilde{u}_{i,j,s} + \tilde{u}_{i,j,s+4})(x_j - v)^\top n_s + (\tilde{u}_{i,j,s} - \tilde{u}_{i,j,s+4})r_s \right) \xi^\top n_s = 0 \,.$$

From $\frac{\partial J}{\partial r_s}$ we obtain

$$r_s = -\frac{\sum_{j=1}^n u_{i,j}^m (\tilde{u}_{i,j,s} - \tilde{u}_{i,j,s+4})(x_j - v)^\top}{\sum_{j=1}^n u_{i,j}^m (\tilde{u}_{i,j,s} + \tilde{u}_{i,j,s+4})} n_s,$$

and substitute this term for r_s in $\frac{\partial J}{\partial v}$. The notations $a_{i,s}$ and $\alpha_{i,s}$ in the theorem lead us to

$$\sum_{s=0}^3 \left(a_{i,s}^\top n_s - \alpha_{i,s} v^\top n_s \right) \xi^\top n_s = 0.$$

For $\xi = n_1$ and $\xi = n_3$ in particular, we get the equations

$$0 = \frac{1}{\sqrt{2}}(a_{i,0}^\top n_0 - \alpha_{i,0} v^\top n_0) + (a_{i,1}^\top n_1 - \alpha_{i,1} v^\top n_1) +$$
$$\frac{1}{\sqrt{2}}(a_{i,2}^\top n_2 - \alpha_{i,2} v^\top n_2),$$

$$0 = \frac{-1}{\sqrt{2}}(a_{i,0}^\top n_0 - \alpha_{i,0} v^\top n_0) + \frac{1}{\sqrt{2}}(a_{i,2}^\top n_2 - \alpha_{i,2} v^\top n_2) +$$
$$(a_{i,3}^\top n_3 - \alpha_{i,3} v^\top n_3).$$

With λ, $\mu \in \mathbb{R}$ and $v = \lambda n_0 + \mu n_2$, we substitute $v^\top n_0 = \lambda$, $v^\top n_1 = \frac{\lambda+\mu}{\sqrt{2}}$, $v^\top n_2 = \mu$ and $v^\top n_3 = \frac{-\lambda+\mu}{\sqrt{2}}$. We rearrange for λ and μ:

$$0 = a_{i,1}^\top n_1 + \frac{a_{i,0}^\top n_0}{\sqrt{2}} + \frac{a_{i,2}^\top n_2}{\sqrt{2}} - \lambda \frac{\alpha_{i,1} + \alpha_{i,0}}{\sqrt{2}} - \mu \frac{\alpha_{i,1} + \alpha_{i,2}}{\sqrt{2}},$$

$$0 = a_{i,3}^\top n_3 - \frac{a_{i,0}^\top n_0}{\sqrt{2}} + \frac{a_{i,2}^\top n_2}{\sqrt{2}} + \lambda \frac{\alpha_{i,3} + \alpha_{i,0}}{\sqrt{2}} - \mu \frac{\alpha_{i,3} + \alpha_{i,2}}{\sqrt{2}}.$$

Multiplying the first equation by $\frac{\alpha_{i,3}+\alpha_{i,0}}{\sqrt{2}}$ (resp. $\frac{\alpha_{i,3}+\alpha_{i,2}}{\sqrt{2}}$) and the second equation by $\frac{\alpha_{i,1}+\alpha_{i,0}}{\sqrt{2}}$ (resp. $\frac{\alpha_{i,1}+\alpha_{i,2}}{\sqrt{2}}$) and adding (resp. subtracting) both resulting equations we obtain

$$\mu = \frac{1}{q}((\sqrt{2}a_{i,1}^\top n_1 + a_{i,0}^\top n_0 + a_{i,2}^\top n_2)(\alpha_{i,3} + \alpha_{i,0}) -$$
$$(\sqrt{2}a_{i,3}^\top n_3 - a_{i,0}^\top n_0 + a_{i,2}^\top n_2)(\alpha_{i,1} + \alpha_{i,0})),$$

$$\lambda = \frac{1}{q}((\sqrt{2}a_{i,1}^\top n_1 + a_{i,0}^\top n_0 + a_{i,2}^\top n_2)(\alpha_{i,3} + \alpha_{i,2}) -$$
$$(\sqrt{2}a_{i,3}^\top n_3 - a_{i,0}^\top n_0 + a_{i,2}^\top n_2)(\alpha_{i,1} + \alpha_{i,2})).$$

The angle φ only appears in the normal vectors n_s. To solve for φ we use $y^\top n_s = (R_s y)^\top n_0$, which is shown easily with the help of the trigonometric addition theorems:

$$(R_s y)^\top n_0$$
$$= \begin{pmatrix} y_1 \cos(\frac{s\pi}{4}) + y_2 \sin(\frac{s\pi}{4}) \\ -y_1 \sin(\frac{s\pi}{4}) + y_2 \cos(\frac{s\pi}{4}) \end{pmatrix}^\top n_0$$
$$= -y_1 (\cos(\varphi) \cos(\frac{s\pi}{4}) - \sin(\frac{s\pi}{4}) \sin(\varphi))$$
$$\quad -y_2 (\sin(\frac{s\pi}{4}) \cos(\varphi) + \sin(\varphi) \cos(\frac{s\pi}{4}))$$
$$= -y_1 \cos(\varphi + \frac{s\pi}{4}) - y_2 \sin(\varphi + \frac{s\pi}{4})$$
$$= (y_1, y_2)^\top n_s.$$

The multiplication of the matrix R_s by a vector x corresponds to a rotation of x by an angle of $\frac{s\pi}{4}$. We substitute $x^\top n_s$ by $(R_s x)^\top$ so that only the normal vector n_0 remains, and we can factor out the angle φ. In this way, we obtain $\lambda = \tilde{l}^\top n_0$ and $\mu = \tilde{m}^\top n_0$, where $\tilde{l}$ and $\tilde{m}$ are defined in the theorem.

To obtain a similar expression for the (half) edge lengths r_s, we transform v into the coordinate system $\{n_s, n_{s+2 \bmod 8}\}$, assuming the coordinate vector (λ_s, μ_s). In the case of $s = 0$ from $v = \lambda n_0 + \mu n_2$ it follows that $(\lambda_0, \mu_0) = (\lambda, \mu)$. A rotation by 45 degrees leads to $(\lambda_1, \mu_1) = (\frac{\lambda+\mu}{\sqrt{2}}, \frac{-\lambda+\mu}{\sqrt{2}})$, $(\lambda_2, \mu_2) = (\mu, -\lambda)$, $(\lambda_3, \mu_3) = (\frac{-\lambda+\mu}{\sqrt{2}}, \frac{-\lambda-\mu}{\sqrt{2}})$, etc. Because n_s and $n_{s+2 \bmod 8}$ are perpendicular $(n_s^\top n_{s+2 \bmod 8} = 0)$ and the normalized vector length of n_s $(n_s^\top n_s = 1)$, we obtain $v^\top n_s = (\lambda_s n_s + \mu_s n_{s+2 \bmod 8}) n_s = \lambda_s \cdot 1 + \mu_s 0 = \lambda_s$. Substituting λ and μ by $\tilde{l}^\top n_0$ and $\tilde{m}^\top n_0$ in the definition of λ_s and μ_s, we get linear combinations of $\tilde{l}^\top n_0$ and $\tilde{m}^\top n_0$ for λ_s $(\lambda_s = \tilde{l}_s n_0$ as defined in the theorem). We also denote r_s in the form $\tilde{r}_s n_0$.

With all these equalities we can rewrite the objective function

$$
\sum_{j=1}^{n} u_{i,j}^m \left(\sum_{s=0}^{7} \tilde{u}_{i,j,s}((x_j - v)^\top n_s + r_{s \bmod 4}) \right)^2
$$

$$
= \sum_{j=1}^{n} u_{i,j}^m \left(\sum_{s=0}^{7} \tilde{u}_{i,j,s}(x_j^\top n_s - (\lambda_s n_s + \mu_s n_{s+2 \bmod 8})^\top n_s + r_{s \bmod 4}) \right)^2
$$

$$
= \sum_{j=1}^{n} u_{i,j}^m \left(\sum_{s=0}^{7} \tilde{u}_{i,j,s}((R_s x_j)^\top n_0 - \lambda_s + \tilde{r}_{s \bmod 4}^\top n_0) \right)^2
$$

$$
= \sum_{j=1}^{n} u_{i,j}^m \left(\sum_{s=0}^{7} \tilde{u}_{i,j,s}(((R_s x_j) - \tilde{l}_s + \tilde{r}_{s \bmod 4})^\top n_0) \right)^2
$$

$$
= n_0^\top \left(\sum_{j=1}^{n} u_{i,j}^m \sum_{s=0}^{7} \tilde{u}_{i,j,s} \right.
$$

$$
\left. (R_s x_j - \tilde{l}_s + \tilde{r}_{s \bmod 4})(R_s x_j - \tilde{l}_s + \tilde{r}_{s \bmod 4})^\top \right) n_0.
$$

Again, this expression is minimal when n_0 is the eigenvector of matrix A with the smallest eigenvalue. From n_0 we obtain φ by applying the atan-function. ■

Because of the convergence problem, we use the same fuzzified minimum functions (5.16) of FCRS for the FC2RS algorithm. Ten steps of the FCM, FCSS and FCRS algorithm provided the initialization for the FC2RS algorithm for the following examples. The FCRS edge lengths r_0 and r_1 were used as the FC2RS edge lengths r_0 and r_2. The remaining edges r_1 and r_3 were initialized by $\sqrt{r_0^2 + r_1^2}$. In this way the FC2RS cluster shape simulates exactly the FCRS rectangle.

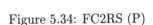

Figure 5.34: FC2RS (P)

The quite flexible clusters of the FC2RS algorithm are appropriate for the approximation of circles or ellipses (see figure 5.34). The exact determination of the parameters is not possible of course, but the algorithm gives good approximate values, which at least allow to distinguish circles from ellipses or rectangles. If the ellipses are well separated, like in figure 5.34, the FC2RS algorithms only needs a few iterations steps if it is initialized by some fuzzy c-means steps.

Figure 5.35: FC2RS (P)   Figure 5.36: FC2RS (P)

If the data set contains elliptic and rectangular shapes simultaneously, the FC2RS algorithm yields better results than the special algorithms for the detection of ellipses (FCQS) or rectangles (FCRS). Figures 5.35 and 5.36 show edges, circles and ellipses detected by the FC2RS algorithm. In this example, the detection of lines even includes their length. Of course, if there are parallel or collinear lines or lines that meet each other at angles of multiples of 45 degrees, several lines are lumped together into one FC2RS cluster.

The cluster contour in figure 5.36 lies slightly inside (not on) the depicted shapes. This is an indication that the minimum function has been chosen too fuzzy. Too many data vectors of the neighbouring edges have been taken into account for estimating the actual edges. In this case, making the minimum function less fuzzy (for example, by using $d = \delta$ instead of $d = 2\delta$ in (5.23)) improves the FC2RS result.

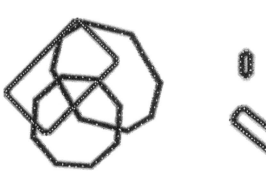

Figure 5.37: FC2RS (P)   Figure 5.38: FC2RS (P)

If the edges are parallel or rotated in steps of 45 degrees, some problems may arise with FC2RS, similar to the FCRS algorithm and an angle of 90 degrees. Then, approximating edges of different shapes by one cluster is possible and leads to a strong local minimum. The data set of figure 5.30 is a good example of such a data set, where FC2RS converges into a local minimum. If the shapes are well separated or have slightly different angles of rotation, the detection is much easier for the FC2RS algorithm, as figure 5.37 and 5.38 demonstrate. For the second figure, the GK and GG algorithms were used for initialization instead of FCSS and FCRS.

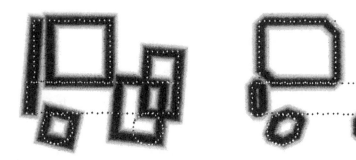

Figure 5.39: FCRS partition (P)   Figure 5.40: FC2RS (P)

Even the detection of rectangles can be improved by FC2RS in some cases, for example if there are circles in the data set which can be better approximated by the FC2RS than the FCRS cluster shape. If the clusters shape is not appropriate for the depicted patterns (as the rectangular shape for the lower left circle in figure 5.39), the detection of the remaining shapes is influenced. A better approximation like in figure 5.40 helps to avoid further misclassifications.

Compared to the crisp minimum functions, the fuzzifications means much more computational costs. Using crisp functions, all but one of the summation terms over the edges evaluate to zero. With fuzzy functions all terms are positive, that means a computational factor of four (resp. eight) for FCRS (resp. FC2RS). The costs of partitioning c clusters can be compared with partitioning $4 \cdot c$ (resp. $8 \cdot c$) line clusters with the AFC algorithm. Sometimes it might be adequate to use the hard minimum functions in the first iteration steps, switching to fuzzy minimum functions later. This saves much time but affects the estimated partition.

At the beginning of the FC(2)RS iteration a bad initialization leads to somewhat chaotic changes in the angle of rotation φ. But the assignment of the data vector to the edges depends on the angle φ. So this effect makes the determination of the correct angle more difficult. Forcing the angle to stay within a range of $[-\frac{\pi}{4}, \frac{\pi}{4}]$ (resp. $[-\frac{\pi}{8}, \frac{\pi}{8}]$) by adding and subtracting multitudes of $\frac{\pi}{2}$ (resp. $\frac{\pi}{4}$) seems to be the obvious thing to do. As long as the angle is evaluated randomly, this constraint prevents the data vectors from being assigned to completely different edges in every iteration step. The performance of the algorithm is not affected, because the cluster's shape is invariant to rotation in steps of 90 degrees (resp. 45 degrees). Problems may arise if the angle of rotation equals $\pm\frac{\pi}{4}$ (resp. $\pm\frac{\pi}{8}$). We must avoid an alternating sequence of angles less than $\frac{\pi}{4}$ and greater than $-\frac{\pi}{4}$. Therefore it is better to restrict the *absolute change* of the angle φ to $\frac{\pi}{4}$ (resp. $\frac{\pi}{8}$).

About convergence of the FC(2)RS algorithm

The proofs of theorems 5.5 and 5.6 are based on the hard minimum functions min_s^h, as introduced by (5.5), whereas the final version of the FC(2)RS algorithm uses modified fuzzy minimum functions. In fact, there is no reason why the objective function should be minimized by using the same rules for protoype update. We had a similar situation with MFCQS, where we only modified the distance measure and *hoped* that minimization would still work (which in fact was not the case with all examples). Now, we want to show that the use of the different versions causes only slight changes in the objective function. Therefore, we believe that the algorithm presented works fine with fuzzy minimum functions.

We denote the distance of a data vector x to all four edges of a rectangle k by d_0, d_1, d_2 and d_3. Using the hard minimum functions, the distance function of theorem 5.5 reduces to $\min\{d_0, d_1, d_2, d_3\}$. The following lemma gives an estimation of the error caused by the fuzzification.

Lemma 5.7 Let $f : \mathbb{R}_+ \to \mathbb{R}_+$ be a strictly increasing function with $f(0) = 0$ and $\eta \in \mathbb{R}_+$. Denote the four fuzzy minimum functions by $\min_s : \mathbb{R}^4 \to [0, 1]$, $(d_0, d_1, d_2, d_3) \mapsto \frac{1}{\sum_{i=0}^{3} \frac{(f(d_s-m)+\eta)^2}{(f(d_i-m)+\eta)^2}}$, with $m = \min\{d_0, d_1, d_2, d_3\}$. Then

$$\left\| \left(\sum_{s=0}^{3} \min_s(d_0, d_1, d_2, d_3) d_s \right) - m \right\| < 3\eta^2.$$

Proof: We introduce the abbreviation D_s for $(f(d_s - m) + \eta)^2$ and $s \in \mathbb{N}^*_{<4}$. This implies

$$\left\| \left(\sum_{s=0}^{3} \min_s(d_0, d_1, d_2, d_3) d_s \right) - m \right\|$$

$$= \left\| \frac{d_0}{\frac{D_0}{D_0} + \frac{D_0}{D_1} + \frac{D_0}{D_2} + \frac{D_0}{D_3}} + \right.$$
$$\left. \frac{d_1}{\frac{D_1}{D_0} + \frac{D_1}{D_1} + \frac{D_1}{D_2} + \frac{D_1}{D_3}} + \ldots - m \right\|$$

$$= \left\| \frac{d_0}{1 + \frac{D_0 D_2 D_3 + D_0 D_1 D_3 + D_0 D_1 D_2}{D_1 D_2 D_3}} + \right.$$
$$\left. \frac{d_1}{1 + \frac{D_1 D_2 D_3 + D_0 D_1 D_3 + D_0 D_1 D_2}{D_0 D_2 D_3}} + \ldots - m \right\|$$

$$= \left\| \frac{d_0 D_1 D_2 D_3 + D_0 d_1 D_2 D_3 + D_0 D_1 d_2 D_3 + D_0 D_1 D_2 d_3}{D_0 D_1 D_2 + D_0 D_1 D_3 + D_0 D_2 D_3 + D_1 D_2 D_3} - m \right\|$$

$$\overset{\star}{=} \left\| \frac{(d_3-d_0) D_0 D_1 D_2 + (d_2-d_0) D_0 D_1 D_3 + (d_1-d_0) D_0 D_2 D_3}{D_0 D_1 D_2 + D_0 D_1 D_3 + D_0 D_2 D_3 + D_1 D_2 D_3} \right\|$$

$$\overset{\star\star}{\leq} \left\| \frac{D_0[(d_3-d_0) D_1 D_2 + (d_2-d_0) D_1 D_3 + (d_1-d_0) D_2 D_3]}{D_1 D_2 D_3} \right\|$$

$$\overset{\star\star\star}{<} \left\| \frac{D_0(3 D_1 D_2 D_3)}{D_1 D_2 D_3} \right\|$$

$$= \|3 D_0\| = 3(f(d_0 - m) + \eta)^2 = 3(f(d_0 - d_0) + \eta)^2$$

$$= 3\eta^2.$$

For equation $\star$ we assume without loss of generality that $d_0 = \min(d_0, d_1, d_2, d_3) = m$. Inequality $\star\star$ results from the fact that $D_0 D_1 D_2$,

$D_0 D_1 D_3$ and $D_0 D_2 D_3 > 0$, inequality $\star \star \star$ from $(d_s - d_0) \leq f(d_s) <$ $f(d_s) + \eta < D_s$ for all $s \in \mathbb{N}^*_{\leq 4}$. (This Lemma can easily be generalized for the case of the FC2RS algorithm). ■

If we choose $f(x) = x$ and $\eta = \varepsilon$ we get an estimation for the error of the objective function caused by choosing fuzzy instead of hard minimum functions (5.14). If the distance between neighbouring data vectors is equal to 1 and ε is choosen by (5.23), the error is $3\frac{9}{25} = 1.08$, i.e. about image resolution. For the final modification (5.16) we use $f(x) = \exp(x)$. Because in the Lemma η is fixed, we consider all $(d_0, d_1, d_2, d_3) \in \mathbb{R}^4$ with the same minimum m. Then, modification (5.16) is equal to the case of $\eta = \varepsilon + m$. In this way, we can see that the error of the fuzzy minimum function increases with m. As mentioned above, the greater the minimum (e.g. the distance of a data vector to the rectangle) is, the fuzzier the minimum function gets. But near the border of the rectangle the minimum m approaches zero. Therefore, we have the case $f(x) = exp(x)$ and $\eta = \varepsilon$ and the Lemma gives the same error-estimation. These observations provide a plausible explanation, that we have to expect only slight differences in the objective function, if we use fuzzy instead of hard minimum functions.

5.4 Computational effort

The FC(2)RS algorithms have a time index of $\tau_{FCRS} = 4.7 \cdot 10^{-4}$ resp. $\tau_{FC2RS} = 9.5 \cdot 10^{-4}$. As we expected, we have approximately twice the costs for FC2RS than for FCRS, because the cluster shape of FC2RS uses twice the number of lines. Compared to the AFC, FCV or GK algorithm, we do not have a factor of four for the time index ($4\tau = 5.8 \cdot 10^{-4} < \tau_{FCRS}$), because the cluster orientation has to be calculated only once for every four lines.

The examples contain 200 to 300 data vectors, figures 5.35 and 5.36 only 70. The number of iteration steps varied from 7 (for figure 5.34) to 120, in some cases even much greater.

Chapter 6

Cluster Estimation Models

In chapters $1 - 5$ we have introduced various models for fuzzy clustering: the basic models fuzzy and possibilistic c-means, models with linear and ellipsoidal prototypes, and models which consider polygonal object boundaries. All these models basically involve the minimization of an objective function like

$$J_{\text{FCM}}(U, V; X) = \sum_{k=1}^{n} \sum_{i=1}^{c} u_{ik}^{m} d_{ik}^{2} \tag{6.1}$$

with the data set $X = \{x_1, \ldots, x_n\} \in \mathbb{R}^p$, the membership matrix $U \in M \subseteq [0, 1]^{c \cdot n}$, the parameters V of the c prototypes, the distance d_{ik} between x_k and the ith prototype, $i = 1, \ldots, c$, $k = 1, \ldots, n$, and the fuzziness parameter $m \in (1, \infty)$.

In the previous chapters the models specified by the objective function (e.g. 6.1,) were minimized using alternating optimization (AO). It was proven by Bezdek [10] that the sequences $\{(U^{(1)}, V^{(1)}), (U^{(2)}, V^{(2)}), \ldots\}$ generated by FCM-AO for fixed $m > 1$ always converge (or contain a subsequence that does) to local minima or saddle points of the objective function J_{FCM}. In many practical examples alternating optimization found almost optimal results with respect to J after a feasible number of iterations. Saddle points seem to be only a theoretical problem and local minima can often be avoided.

There are, however, many other methods to minimize clustering models (objective functions), e.g. hybrid schemes (relaxation) [47], genetic algorithms [1], reformulation [42], and artificial life methods [86]. Therefore, it is important to clearly distinguish between clustering models like FCM

157

and the algorithms used to minimize the respective objective function like FCM-AO.

In this chapter we consider the general architecture of the alternating optimization algorithm, but abandon the purpose of minimizing objective functions. By generalizing the optimization algorithm itself we obtain the *alternating cluster estimation (ACE)* [91], which has a much higher flexibility than AO and leads in some cases to a considerable improvement of the clustering results.

6.1   AO membership functions

Alternating optimization was already introduced in chapter 1 as a basic clustering algorithm. Let us briefly recall how AO works: the prototypes are randomly initialized as $V^{(0)}$. At each optimization step t the partition $U^{(t)}$ and the prototypes $V^{(t)}$ are updated. The algorithm terminates when the maximum number of iterations is reached, $t = t_{\max}$, or when successive approximations to the prototypes have stabilized, $\|V^{(t)} - V^{(t-1)}\|_\varepsilon < \varepsilon_V$, where $\varepsilon_V \in \mathbb{R}^+\backslash\{0\}$ and $\|.\|_\varepsilon$ is any appropriate norm. A similar algorithm AO' is obtained by exchanging U and V in AO. In AO' the partition $U^{(0)}$ is randomly initialized and AO' terminates if $\|U^{(t)} - U^{(t-1)}\|_\varepsilon < \varepsilon_U$, $\varepsilon_U \in \mathbb{R}^+\backslash\{0\}$. Here we focus on AO because it requires less comparisons and storage than AO'.

In the previous chapters we presented AO solutions of fuzzy and possibilistic clustering models with different prototypes. The equations to update the partitions U and the prototypes V were the first order necessary conditions for local extrema of the respective objective functions. To update the memberships we obtained, e.g. for the fuzzy c-means model

$$u_{ik} = 1 \Big/ \sum_{j=1}^{c} \left(\frac{\|x_k - v_i\|_A}{\|x_k - v_j\|_A} \right)^{\frac{2}{m-1}}, \quad i = 1,\ldots,c, \quad k = 1,\ldots,n, \quad (6.2)$$

and for the possibilistic c-means model

$$u_{ik} = \frac{1}{1 + \left(\frac{\|x_k - v_i\|_A}{\sqrt{\eta_i}} \right)^{\frac{2}{m-1}}}, \quad i = 1,\ldots,c, \quad k = 1,\ldots,n. \quad (6.3)$$

For both FCM and PCM we obtained the prototype update equation

$$v_i = \frac{\sum\limits_{k=1}^{n} u_{ik}^m \, x_k}{\sum\limits_{k=1}^{n} u_{ik}^m}, \quad i = 1,\ldots,c. \quad (6.4)$$

The memberships u_{ik} calculated by AO using (6.2) or (6.3), e.g., can be interpreted as discrete samples of continuous membership functions $\mu_i :$ $\mathbb{R}^p \to [0,1]$ where

$$\mu_i(x_k) = u_{ik}, \quad i = 1, \ldots, c, \quad k = 1, \ldots, n. \tag{6.5}$$

Appropriate membership functions μ_i can be formally obtained by replacing x_k with x in the partition update equations and choosing reasonable definitions for the membership values at the cluster centres. The resulting continuous membership function $\mu_i(x)$, $i = 1, \ldots, c$, for FCM-AO (6.2), is

$$\mu_i(x) = \begin{cases} 1 \left/ \sum_{j=1}^{c} \left(\frac{\|x - v_i\|_A}{\|x - v_j\|_A} \right)^{\frac{2}{m-1}} \right. & \text{for } x \in \mathbb{R}^p \setminus V, \\ 1 & \text{for } x = v_i, \\ 0 & \text{for } x \in V \setminus \{v_i\}, \end{cases} \tag{6.6}$$

and for PCM-AO (6.3), $\eta_i > 0 \; \forall i$,

$$\mu_i(x) = \frac{1}{1 + \left(\frac{\|x - v_i\|_A}{\sqrt{\eta_i}} \right)^{\frac{2}{m-1}}}, \quad i = 1, \ldots, c. \tag{6.7}$$

Figure 6.1 shows the memberships u_{ik} obtained with FCM-AO (6.2) for $X = \{0, 1, \ldots, 50\}$, $c = 3$, $m = 2$, $t_{\max} = 100$, $\|V\|_\varepsilon = \max_{i=1,\ldots,c;\, l=1,\ldots,p} \{v_i^{(l)}\}$, and $\varepsilon_V = 10^{-10}$. The memberships u_{2k} decrease for $v_2 < x < v_3$ (arrow A), and increase again for $x > v_3$ (arrow B). The associated membership function μ_2 (solid line in figure 6.1) is non-convex as a fuzzy set [104], i.e. there are non-convex α cuts $\mu_2^{\geq \alpha}(x)$. Membership functions $\mu_i(x)$ from FCM-AO (6.6) are non-convex fuzzy sets.

6.2 ACE membership functions

In fuzzy rule based systems like fuzzy controllers [29] fuzzy sets are assigned linguistic labels like "low", "medium", and "high". The semantic meaning of these labels requires the membership functions to be convex or even monotonous. Also fuzzy numbers [48] like "about 100" are frequently used in fuzzy rule based systems. Fuzzy numbers are by definition convex fuzzy sets. Convexity of the fuzzy sets is an important requirement in these rule based systems. Since the FCM-AO membership functions are non-convex, we want to modify the AO algorithm. With the modified algorithm it should be possible to generate convex and other membership functions.

In general, the continuous membership functions obtained from AO are restricted to the special shape determined by the update equation

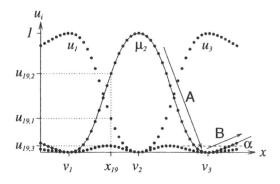

Figure 6.1: Memberships and membership functions from FCM-AO

$U^{(t)}(V^{(t-1)}, X, \ldots)$, which is derived from the objective function for the model (figure 6.2 left). Designers of fuzzy rule based systems, however, use a wide variety of membership function shapes such as Gaussian, triangular, or trapezoidal, which are implemented in modern fuzzy design tools like SieFuzzy®. The modified clustering algorithm should therefore be able to generate arbitrary membership function shapes.

To achieve this we abandon the objective function model and present a more general model which is defined by the *architecture* of the AO algorithm and by *user specified* equations to update U and V (figure 6.2 right). The update equations chosen by the user do not necessarily reflect an *optimization* of a particular objective function. When the user selects update equations that do not correspond to an objective function model, clusters and cluster centres are *estimated* by alternatingly updating partitions and prototypes. Therefore, we call this generalized clustering algorithm *alternating cluster estimation (ACE)*.

Some appropriate candidates for the user selectable membership functions in the ACE model are shown in figure 6.3. In the practical implementation the user can select membership function families with the "toolbar" in the partition builder (figure 6.4).

Notice that the conventional (objective function defined) clustering algorithms can be represented as instances of the more general ACE model. For FCM-AO we select FCM membership functions (6.2) with the Euclidean norm and $m = 2$ in the partition builder. To obtain PCM-AO we change the selection in the partition builder to Cauchy membership functions (6.3)

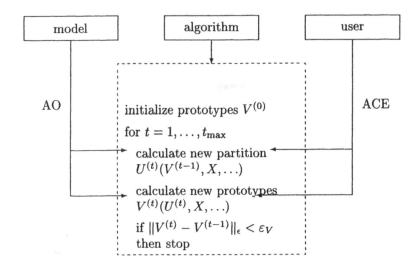

Figure 6.2: Alternating optimization and alternating cluster estimation

with the Euclidean norm, $m = 2$, and

$$\eta_i = \frac{\sum\limits_{k=1}^{n} u_{ik}^m \|x_k - v_i\|^2}{\sum\limits_{k=1}^{n} u_{ik}^m}, \quad i = 1, \ldots, c. \tag{6.8}$$

The other selections from the toolbar in figure 6.4 lead to ACE algorithms which can not easily be associated with clustering algorithms specified by objective functions. In the following section we describe an example for the family of these "pure" ACE algorithms.

6.3 Hyperconic clustering (dancing cones)

For hyperconic clustering we use the ACE algorithm and select hyperconic membership functions defined as

$$\mu_i(x) = \begin{cases} 1 - \frac{d_i(x)}{r_i} & \text{for } d_i(x) \le r_i \\ 0 & \text{otherwise,} \end{cases} \tag{6.9}$$

in the partition builder, where $d_i(x)$ is the distance between the point x and the ith cluster. For clusters specified by the centre v_i we use the Euclidean

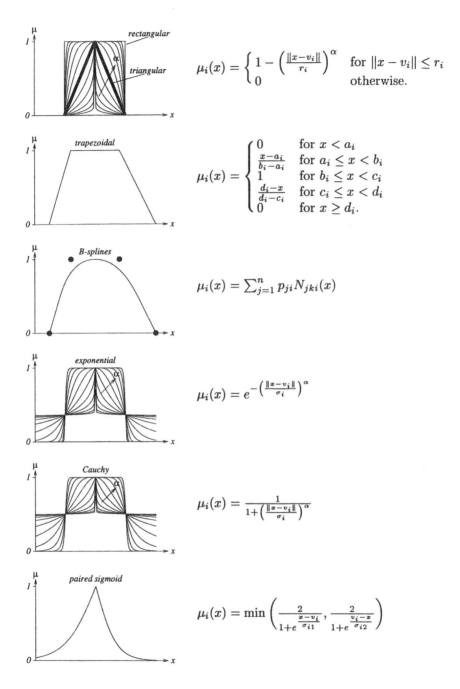

Figure 6.3: Some ACE membership functions and their equations

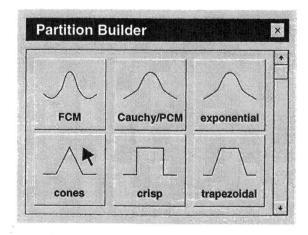

Figure 6.4: ACE membership function toolbar in the partition builder

distance $d_i(x) = \|x - v_i\|$ here. For the choice of the hypercone radii r_i, $i = 1, \ldots, c$, we consider the hyperbox $H(X)$ spanned by X (Fig. 6.5). If we denote the side lengths of $H(X)$ as s_l, $l = 1, \ldots, p$, a good choice for r_i seems to be

$$r_i = \frac{\sqrt{\sum_{l=1}^{p} s_l^2}}{c + 1}. \tag{6.10}$$

In this case a row of c equal sized "kissing" hypercones exactly spans the diagonal through $H(X)$. As ACE proceeds to alternate between estimates of $V^{(t)}$, $t \in \mathbb{N}_0$, the membership functions (the cones) change *only* by the location of their centres (the current estimate of V). Thus, the cones are fixed in height and base radius, but "dance around" in $\mathbb{R}^p$ as $V^{(t-1)} \to V^{(t)}$, always being used to obtain $U^{(t)}$ from $V^{(t-1)}$ by evaluation of (6.9) with $(X, V^{(t-1)})$. We call this instance of ACE *hyperconic clustering* or *dancing cones (DC)*. Notice the difference between the supports of DC, FCM-AO, and PCM-AO membership functions. With the global FCM-AO and PCM-AO membership functions the cluster centres are also influenced by very distant data points. With the local DC membership functions each cluster centre v_i is only affected by the data points in the local hyperball around v_i with the radius r_i.

For a comparison we apply FCM-AO, PCM-AO, and DC (figure 6.6) to six simple clustering problems. For the first four we fill two distinct square

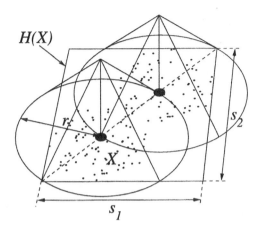

Figure 6.5: Choice of the cluster radius

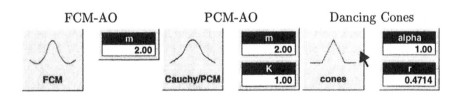

Figure 6.6: Examples of section 6.3

areas a_1 and a_2 in $[0,1]^2$ with n_1 and n_2 random points, respectively. We denote the respective point densities as $d_1 = n_1/a_1$ and $d_2 = n_2/a_2$ and obtain the data sets $X_1, \ldots, X_4$ according to the following table:

X	n_1	n_2	a_1	a_2	d_1	d_2	
X_1	100	100	$\left(\frac{1}{2}\right)^2$	$\left(\frac{1}{2}\right)^2$	400	400	$*_1 = *_2$
X_2	100	100	$\left(\frac{2}{3}\right)^2$	$\left(\frac{1}{3}\right)^2$	225	900	$n_1 = n_2$
X_3	160	40	$\left(\frac{1}{2}\right)^2$	$\left(\frac{1}{2}\right)^2$	640	160	$a_1 = a_2$
X_4	160	40	$\left(\frac{2}{3}\right)^2$	$\left(\frac{1}{3}\right)^2$	360	360	$d_1 = d_2$

We use cluster centres randomly initialized as $v_i^{(0)} \in [0,1]^2$, $i = 1, \ldots, c$, the parameters $c = 2$, $t_{\max} = 100$, $\|V\|_\varepsilon = \max_{i=1,\ldots,c;\, l=1,\ldots,p}\{v_i^{(l)}\}$, and $\varepsilon_V = 10^{-10}$. The data sets and the cluster centres obtained by the three algorithms are shown in figure 6.7. FCM-AO and DC produce good cluster centres in all four experiments, while PCM-AO generates two equal cluster centres for X_2, X_3, and X_4. PCM-AO seems to interpret the cluster with lower density or with less points as noise and does not recognize it as a cluster. Also notice that for X_4, DC produces a visually superior result to FCM-AO's estimate of the upper cluster centre.

The last two experiments use noisy data sets (figure 6.8). X_5 contains $n_1 = n_2 = 50$ random points in two small areas $a_1 = a_2 = (1/10)^2$ and 100 additional points randomly distributed over $[0,1]^2$. PCM-AO and DC detect and represent the two classes correctly, while FCM-AO is distracted by the background noise. In X_6 the original data set X_1 was used, but one severe outlier $x_K = (30, 30)$ was added to it, which is not visible in figure 6.8. The results of PCM-AO and DC are not affected by the outlier, while FCM-AO moves both centres towards the outlier, the upper centre not visible in $[0,1]^2$, as indicated by the arrow.

In our experiments FCM-AO performed well, except for the data sets containing severe background noise or outliers (X_5, X_6). PCM-AO handled noise and outliers well, but sometimes it did not recognize clusters with lower density or less points ($X_2, \ldots, X_4$). The dancing cones algorithm in the ACE environment with hyperconic membership functions produced acceptable results in all six cases.

6.4 Prototype defuzzification

We obtained the ACE model with used defined membership functions by customizing the equations to update the partitions in the AO algorithm. Now we customize the equations to update the *prototypes*. In FCM-AO,

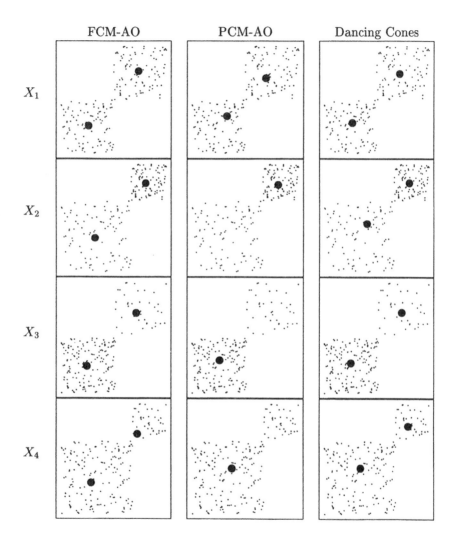

Figure 6.7: Clustering results for $X_1, \ldots, X_4$ in example 6.3

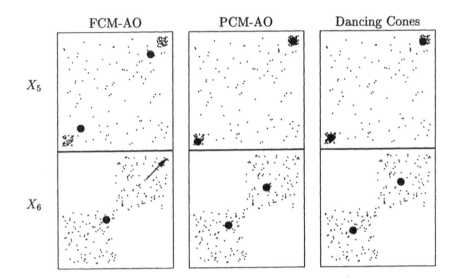

Figure 6.8: Clustering results for X_5 and X_6 in example 6.3

PCM-AO, and also in the dancing cones algorithm the prototypes are updated using (6.4)

$$v_i = \frac{\sum\limits_{k=1}^{n} u_{ik}^m x_k}{\sum\limits_{k=1}^{n} u_{ik}^m}, \quad i = 1, \dots, c. \tag{6.11}$$

Given a fuzzy set described by the memberships u_{ik} of the data points x_k
the prototype update equation computes a crisp centre point v_i which is
significant with respect to the fuzzy set. The update equation therefore
converts fuzzy information into crisp – it is a so-called *defuzzification*. In
a different context, Filev and Yager [34] defined the *basic defuzzification
distribution (BADD)* as

$$d_{\text{BADD}}(U_i, X, \gamma) = \frac{\sum\limits_{k=1}^{n} u_{ik}^\gamma x_k}{\sum\limits_{k=1}^{n} u_{ik}^\gamma}, \quad \gamma \in \mathbb{R}^+ \backslash \{0\}, \tag{6.12}$$

which is equivalent to the FCM/PCM-AO prototype update equation (6.11)
for $\gamma = m$. Depending on the exponent γ BADD has some interesting
special cases:

- For $\gamma = 1$ BADD becomes the *centre of gravity (COG)* defuzzification.

$$d_{\text{BADD}}(U_i, X, 1) = d_{\text{COG}}(U_i, X). \qquad (6.13)$$

- For an arbitrary set of nonzero memberships $U_i^{\heartsuit} = \{u_{i1}, \ldots, u_{in}\}$, $u_{ik} \in (0, 1]$, $k = 1, \ldots, n$, we have

$$\lim_{\gamma \to 0} \{d_{\text{BADD}}(U_i^{\heartsuit}, X, \gamma)\} = \lim_{\gamma \to 0} \left\{ \frac{\sum_{k=1}^n u_{ik}^{\gamma} x_k}{\sum_{k=1}^n u_{ik}^{\gamma}} \right\} = \frac{1}{n} \sum_{k=1}^n x_k = \bar{x},$$
$$(6.14)$$

which is the mean of the data set X.

- We assume $\hat{u} = \max\{u_{i1}, \ldots, u_{in}\} > 0$, denote $u_{ik}' = u_{ik}/\hat{u}$, $k = 1, \ldots, n$, and obtain, for $\gamma \to \infty$

$$\lim_{\gamma \to \infty} \{d_{\text{BADD}}(U_i, X, \gamma)\} = \lim_{\gamma \to \infty} \left\{ \frac{\sum_{k=1}^n u_{ik}^{\gamma} x_k}{\sum_{k=1}^n u_{ik}^{\gamma}} \right\}$$

$$= \lim_{\gamma \to \infty} \left\{ \frac{\sum_{k=1}^n (u_{ik}')^{\gamma} x_k}{\sum_{k=1}^n (u_{ik}')^{\gamma}} \right\} = \frac{\sum_{u_{ik}'=1} x_k}{\sum_{u_{ik}'=1} 1}$$

$$\dot{=} \frac{\sum_{u_{ik}=\hat{u}} x_k}{\sum_{u_{ik}=\hat{u}} 1} = d_{\text{MOM}}(U_i, X), \qquad (6.15)$$

which is known in the fuzzy literature as the *mean of maxima (MOM)* defuzzification.

In the ACE model we allow the user to specify the parameter $\gamma \in \mathbb{R}^+ \backslash \{0\}$ for the prototype estimation,

$$v_i = d_{\text{BADD}}(U_i, X, \gamma), \qquad (6.16)$$

which includes the special cases COG ($\gamma = 1$) and the prototype equations for FCM-AO and PCM-AO ($\gamma = m$). Moreover, we can extend the prototype toolbar by other families of defuzzification methods (see the survey in [84]). Here are some promising candidates.

- semi-linear defuzzification (SLIDE)
  If we denote $t_{\text{BADD}}(u_{ik}, \gamma) = u_{ik}^{\gamma}$ and $T_{\text{BADD}}(U_i, \gamma) = \{t_{\text{BADD}}(u_{i1}, \gamma), \ldots, t_{\text{BADD}}(u_{in}, \gamma)\}$ as the *BADD transformation* of u_{ik} and U_i, respectively, we can interpret BADD as the COG with transformed weights.

$$d_{\text{BADD}}(U_i, X, \gamma) = d_{\text{COG}}(T_{\text{BADD}}(U_i, \gamma), X). \qquad (6.17)$$

Yager and Filev [102] approximated the BADD transformation by the *semi-linear transformation*

$$t_{\text{SLIDE}}(u_{ik}, \alpha, \beta) = \begin{cases} u_{ik} & \text{if } u_{ik} \geq \alpha \\ (1 - \beta) u_{ik} & \text{if } w < \alpha, \end{cases} \tag{6.18}$$

$\alpha, \beta \in [0,1]$, and obtained the *semi-linear defuzzification (SLIDE)*

$$d_{\text{SLIDE}}(U_i, X, \alpha, \beta) = d_{\text{COG}}(T_{\text{SLIDE}}(U_i, \alpha, \beta), X). \tag{6.19}$$

- modified semi-linear defuzzification (MSLIDE)
  Using a modification of the semi-linear transformation, Yager and Filev [102] defined the *modified semi-linear defuzzification (MSLIDE)* as

$$d_{\text{MSLIDE}}(U_i, X, \beta) = \beta \cdot d_{\text{MOM}}(U_i, X) + (1 - \beta) \cdot d_{\text{COG}}(U_i, X). \tag{6.20}$$

MSLIDE contains the special cases MOM and COG.

The BADD and SLIDE transformations are monotonically increasing functions $t : [0,1] \to [0,1]$ satisfying $t(0) = 0$ and $t(1) = 1$. Other transformations with these properties might also lead to interesting defuzzification methods.

- centre of largest area (COG*)
  Pfluger *et al.* [82] consider U_i^*, the convex fuzzy subset of U_i with the largest cardinality $\sum_{k=1}^n u_{ik}^*$, and define the *centre of largest area* (COG*) defuzzification as the centroid of U_i^*.

$$d_{\text{COG}^*}(U_i, X) = d_{\text{COG}}(U_i^*, X). \tag{6.21}$$

In a similar manner we define

$$
\begin{aligned}
d_{\text{BADD}^*}(U_i, X, \gamma) &= d_{\text{COG}}(T_{\text{BADD}}(U_i^*, \gamma), X), & (6.22) \\
d_{\text{SLIDE}^*}(U_i, X, \alpha, \beta) &= d_{\text{COG}}(T_{\text{SLIDE}}(U_i^*, \alpha, \beta), X), & (6.23) \\
d_{\text{MSLIDE}^*}(U_i, X, \beta) &= d_{\text{COG}}(T_{\text{MSLIDE}}(U_i^*, \beta), X). & (6.24)
\end{aligned}
$$

These and other candidates can be arranged in the ACE prototype toolbar, as shown in figure 6.9.

In the following example we couple the FCM membership function (6.2) using the Euclidean norm $\|x\| = \sqrt{x^T x}$ and $m = 2$ in the partition builder with BADD formula (6.12) in the prototype builder, so that $v_i = d_{\text{BADD}}(U_i, X, \gamma)$, $i = 1, \ldots, c$, where $\gamma \in \mathbb{R}^+ \backslash \{0\}$ is the experimental variable (figure 6.10). For $X \subset \mathbb{R}^2$, we initialize the cluster centres as random vectors

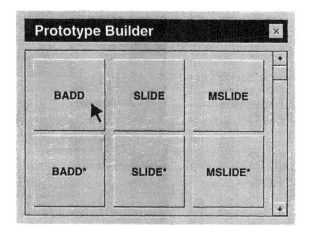

Figure 6.9: User defined prototype builders in ACE

Figure 6.10: ACE configuration for example 6.4

$v_i^{(0)} \in [0,1]^2$, $i = 1, \ldots, c$, and calculate $c = 4$ clusters by iterating ACE with $t_{max} = 100$, $\|V\|_\varepsilon = \max_{i=1,\ldots,c;\, l=1,\ldots,p}\{v_i^{(l)}\}$, and $\varepsilon_V = 10^{-10}$.

We apply this clustering algorithm to the function approximation problem described by Narazaki and Ralescu in [77]. The function

$$y = g(x) = 0.2 + 0.8(x + 0.7\sin(2\pi x)), \quad x \in [0,1], \tag{6.25}$$

is used to generate the 21 equidistant samples

$$X = \{(0, g(0)), (.05, g(.05)), (.1, g(.1)), \ldots, (1, g(1))\}. \tag{6.26}$$

figure 6.11 shows the points $\{(x_1, g(x_1)), \ldots, (x_{21}, g(x_{21}))\}$ (small dots) and the cluster centres $\{(v_1, g(v_1)), \ldots, (v_c, g(v_c))\}$) obtained for various values of γ (big dots).

For $\gamma = m = 2$ the algorithm is FCM-AO and generates a set of cluster centres which we consider a good representation of the data set X. Decreasing γ towards zero moves the cluster centres towards the mean $\bar{x}$ of X. This observation corroborates the limit at (6.14). When γ is continuously increased, the cluster centres v_i move closer to some of the data points $x_k \in X$ (see $\gamma = 8$). As $\gamma \to \infty$ the BADD defuzzification becomes MOM (6.15). MOM simply determines the cluster centres v_i, $i = 1, \ldots, c$, as the data points x_{k_i} which are the closest to the initial values $v_i^{(0)}$.

To avoid this trivial solution we can change the value of γ after each estimation step by, e.g. $\gamma(t) = 2 + t/10$. During the first few steps this method is similar to FCM-AO and is very much like ascending FLVQ [15], because $\gamma \approx 2$. For increasing values of t the algorithm seems to change from a global to a local view, because it moves the cluster centres towards the nearest data points. The resulting cluster centres approximate a set of data points which can be interpreted as a representative subset of X [70].

6.5 ACE for higher-order prototypes

ACE can also be used for higher order prototypes like lines or elliptotypes. To obtain ACE for higher order prototypes we replace the membership update equation with a user-defined membership function and the prototype update equation with a user-defined prototype function in the respective AO algorithm. If we do this with FCE-AO, we obtain ACE for elliptotypes (ACE-E or ACE^2) [88].

In the following experiments we again consider the approximation of the function (6.25) described in the previous section. This time we want to generate a first order Takagi-Sugeno (TS) system [98] to approximate the function $g(x)$ at (6.25). TS systems are sets of c fuzzy rules of the form

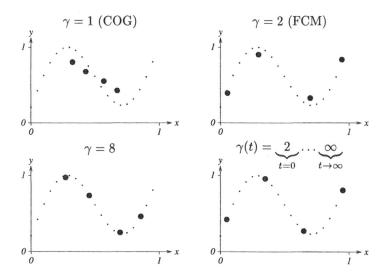

Figure 6.11: Clustering results using different parameters γ in example 6.4

$$\text{IF } T\left(\mu_{i,1}(\vec{x}), \mu_{i,2}(\vec{x}), \ldots, \mu_{i,n}(\vec{x})\right) \text{ THEN } \vec{y} = \vec{f}_i(\vec{x}) \quad 1 \leq i \leq c,$$

where $\vec{x}$ and $\vec{y}$ are the input and output vectors, respectively, $\mu_{i,j}(\vec{x})$ are the fuzzy sets partitioning the input space, T is a t-norm [94], and $\vec{f}_i(\vec{x})$ are functions mapping the input space to the output space. If the functions $\vec{f}_i(\vec{x})$ are linear functions, we call the fuzzy system a *first order TS system*. For one-dimensional input and one-dimensional output (as in this experiment) each linear function can be written as

$$f_i(x) = m_i(x - x_i) + y_i \qquad (6.27)$$

and can be interpreted as a line through the point (x_i, y_i) with the slope m_i. In [87, 88] it was shown that smoothly matched approximations can only be achieved when the first order TS systems are unnormalized. The output of an unnormalized TS system is evaluated as

$$y(x) = \sum_{i=1}^{c} \mu_i(x) \cdot f_i(x). \qquad (6.28)$$

With unnormalized first order TS systems, the best approximations were obtained in [87, 88] using piecewise quadratic membership functions

$$\mu_i(x) = \alpha_i x^2 + \beta_i x + \gamma_i. \qquad (6.29)$$

The polynomial coefficients α_i, β_i, and γ_i are the solutions of the linear equation system representing the following conditions:

1. The output curve $y(x)$ exactly matches the RHS values y_i and the slopes m_i at the rule centres x_1 and x_2, and

2. the membership functions $\mu_i(x)$ are equal to one at the rule centres x_i.

After some conversions we obtain

$$
\begin{aligned}
\alpha_i &= \Big[-(x_i - x_j)\big(m_j(y_i - m_i x_i) - m_i(y_j - m_j x_j)\big) \\
&\quad -2y_j\big((y_i - m_j x_i) - (y_j - m_j x_j)\big) \Big] \\
&\quad / \Big[(x_i - x_j)^3 \big(m_i(y_j - m_j x_j) - m_j(y_i - m_i x_i)\big) \Big], \quad (6.30)
\end{aligned}
$$

$$
\begin{aligned}
\beta_i &= \Big[2(x_i - x_j)x_j\big(m_j(y_i - m_i x_i) - m_i(y_j - m_j x_j)\big) \\
&\quad + 2y_j(x_i + x_j)\big((y_i - m_j x_i) - (y_j - m_j x_j)\big) \Big] \\
&\quad / \Big[(x_i - x_j)^3 \big(m_i(y_j - m_j x_j) - m_j(y_i - m_i x_i)\big) \Big], \quad (6.31)
\end{aligned}
$$

$$
\begin{aligned}
\gamma_i &= \Big[-(x_i - x_j)x_j^2\big(m_j(y_i - m_i x_i) - m_i(y_j - m_j x_j)\big) \\
&\quad -2y_j x_i x_j\big((y_i - m_j x_i) - (y_j - m_j x_j)\big) \Big] \\
&\quad / \Big[(x_i - x_j)^3 \big(m_i(y_j - m_j x_j) - m_j(y_i - m_i x_i)\big) \Big], \quad (6.32)
\end{aligned}
$$

where the index j refers to the rule with the membership function $\mu_j(x)$ which is the closest neighbour of $\mu_i(x)$.

The parameters m_i, x_i, and y_i for the LHS membership functions $\mu_i(x)$ (6.29) with (6.30) – (6.32) and for the RHS functions $f_i(x)$ (6.27) can be computed from the parameters of elliptotypes determined by clustering. Elliptotypes (compare with chapter 3) are fuzzy sets whose α cuts are ellipsoids. Each elliptotype is specified by the point v_i and the directions δ_{ij}, $j = 1, \ldots, q$, $1 \leq q \leq p$. The elliptotype "centre" v_i can be used as the RHS point (x_i, y_i). In the two-dimensional case ($p = 2$) we have $1 \leq q < p = 2 \Rightarrow q = 1$, i.e. each elliptotype has only one direction vector δ_i. This direction vector can be used to compute the RHS slopes

$$
m_i = \delta_i^{(y)} / \delta_i^{(x)}, \quad (6.33)
$$

where $\delta_i^{(x)}$ and $\delta_i^{(y)}$ are the components of δ_i in x and y direction, respectively.

To obtain the elliptotypes, i.e. the cluster centres v_i and directions δ_i, we use (A) the conventional FCE-AO algorithm and (B) an instance of ACE^2. For the best model accuracy we could select quadratic membership functions from the ACE^2 partition builder, but then we would have to compute the polynomial coefficients $\alpha_i, \ldots, \gamma_i$ using (6.30) – (6.32) in each update cycle. We avoid this computational load if we restrict ourselves to the simpler ACE^2 instance with hyperconic membership functions as defined in (6.9), bearing in mind that this decreases the expected accuracy of our fuzzy model.

For the hyperconic membership function at (6.9) we need the distance between the point $x \in \mathbb{R}^p$, and the ith elliptotype specified by v_i and δ_i, $i = 1, \ldots, c$. This distance can be computed as

$$d_i(x) = \sqrt{\|x - v_i\|_A^2 - \alpha \cdot \sum_{j=1}^{q} ((x - v_i)^T A \, \delta_{ij})^2}, \qquad (6.34)$$

where $\alpha \in [0, 1]$ is a geometry parameter specifying the ratio between the axes of the α cut ellipsoids. The ellipsoids become hyperballs for $\alpha = 0$ and hypercylinders for $\alpha = 1$. Notice that the hyperconic membership function at (6.9) with the elliptotype distance function $d_i(x)$ at (6.34) defines a *stretched* hypercone for $\alpha > 0$ and a hypercylinder for $\alpha = 1$. For simplicity we will, however, stick to the simple term hypercone here.

In our ACE^2 instance we update the prototypes in the same way as in FCE-AO. The elliptotype centres v_i are determined by BADD defuzzification, $\gamma = m$, and the elliptotype directions δ_{ij} are computed as the unit eigenvectors of the j largest eigenvalues of the within cluster fuzzy scatter matrices

$$S_i = \sum_{k=1}^{n} u_{ik}^m (x_k - v_i)(x_k - v_i)^T, \quad i = 1, \ldots, c. \qquad (6.35)$$

For both FCE-AO and ACE^2 with hypercones we use the parameters $t_{\max} = 100$, $m = 2$, $q = 1$, $\|x\|_A = \sqrt{x^2}$, and $\alpha = 0.999$. We choose a high value of α to move the cluster centres close to the graph of the function. For ACE^2 we select hyperconic membership functions, $r_i = 1/3$, in the partition builder, and the BADD defuzzification, $\gamma = m = 2$, in the prototype builder (figure 6.12). In order to get a better initialization for ACE^2 we use FCM membership functions for the first 10 iterations. The data set X (6.26) with the 21 samples from $g(x)$ is clustered with FCE-AO and hyperconic ACE^2 with various numbers of clusters c. The results v_i and δ_i, $i = 1, \ldots, c$, are used to compute

1. the coefficients α_i, β_i, γ_i (6.30) – (6.32) of the quadratic LHS membership functions (6.29) and

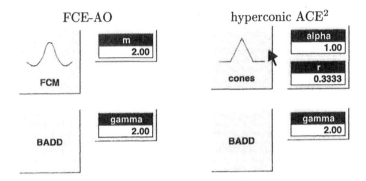

Figure 6.12: Clustering algorithms of section 6.5

2. the points (x_i, y_i) and the slopes m_i for the RHS functions $f_i(x)$
(6.27).

The outputs of the resulting unnormalized first order TS systems with polynomial membership functions are compared with the values of the original
function $g(x)$ (6.25). As a quantitative measure of approximation quality
the resubstitution training error E_1 on the input data and the recall error
E_2 on 101 equally spaced points in $[0, 1]$ are used.

$$E_1 = 100 \times \frac{1}{21} \sum_{i=0}^{20} \frac{|y_i - g(x_i)|}{g(x_i)}\%, \quad E_2 = 100 \times \frac{1}{101} \sum_{i=0}^{100} \frac{|y_i - g(x_i)|}{g(x_i)}\%.$$
(6.36)

Table 6.1 shows the errors obtained with the first order TS systems generated with FCE-AO and hyperconic ACE^2:

Table 6.1 Resubstitution and recall errors

c	FCE-AO		ACE^2	
	E_1	E_2	E_1	E_2
2	28.4%	28.7%	42.3%	34.7%
4	11.9%	11.4%	**5.01%**	**4.04%**
5	5.01%	4.98%	**3.32%**	**2.85%**
11	3.85%	3.5%	**3.63%**	**3.38%**

The first order TS systems generated by ACE^2 yield a lower approximation error than those generated with FCE-AO in most cases (bold numbers).

The approximation with four rules is shown in figure 6.13. The fuzzy system obtained with ACE^2 (right view) leads to a visually more accurate approximation than the fuzzy system obtained with FCE-AO (left view).

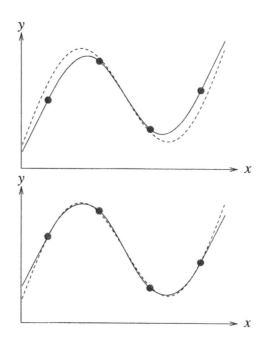

Figure 6.13: Function approximation with FCE-AO and ACE^2, $c = 4S$

When we use only two rules for the approximation with polynomial membership functions, FCE-AO yields lower error values E_1 and E_2 than ACE^2 (table 6.1, first row). E_1 and E_2, however, take into account not only the interpolation, but also the extrapolation error, because they are computed for samples x_i in the whole interval $[0, 1]$. Figure 6.14 shows that the interpolation error of the fuzzy systems obtained with FCE-AO (shaded area in the left view) is much higher than the interpolation error with ACE^2 (shaded area in the right view). Thus, considering only the interpolation error, hyperconic ACE^2 also leads to the fuzzy system with the best approximation for two rules.

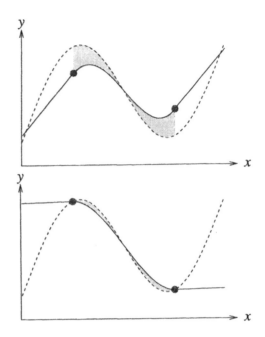

Figure 6.14: Interpolation error for FCE-AO (left) and ACE^2 (right), $c = 2$

6.6 Acceleration of the Clustering Process

ACE allows the choice of computationally inexpensive partition and prototype functions. For example, the computation of the memberships in the
dancing cones model (6.9) is much faster than the computation of the FCM
memberships (6.6). If we think about data mining applications [33], where
Terabytes of data have to be processed, even the faster ACE instances
might lead to inacceptable computation times. To accelerate the clustering
process we present two different approaches:

1. The data set can be partitioned into disjoint blocks or segments. Clustering (or single steps of the clustering algorithm) can then be done
   using one of these blocks or segments instead of the whole data set.
   This leads to an acceleration that directly depends on the size of the
   blocks or segments. The error caused by this approximation is low
   if the blocks or segments are representative for the whole data set.
   This *fast alternating cluster estimation (FACE)* [92] is introduced in
   section 6.6.1.

2. In many applications the partition of the input space should be regular
   and uniform. This is often the case in the identification step of fuzzy
   control applications [29], where equidistant fuzzy sets like "very low",
   "low", "medium", "high", and "very high" are used. These fuzzy
   sets in the input space represent the input space components of the
   cluster centres. If these are defined a priori, only the output space
   components of the cluster centres have to be computed during run
   time. This of course leads to an acceleration of the clustering process.
   If the distances between data and clusters are measured in the input
   space only, this *regular alternating cluster estimation (rACE)* [90]
   can even be reduced to a *single step* algorithm, i.e. prototypes and
   partitions have to be computed *only once*. rACE is presented in more
   detail in section 6.6.2.

6.6.1   Fast Alternating Cluster Estimation (FACE)

To accelerate clustering we can use a *selective* approach. If the data possess
a clear cluster structure, a good approximation of the data set can be
obtained when, for example, only every second point is used. More generally
we can use only every $(n/l)^{th}$ point, i.e. instead of the whole data set $X =
\{x_1, \ldots, x_n\}$ we consider only one of the blocks Y_j, $j = 1, \ldots, n/l$, shown in
figure 6.15. The processed data set is then reduced from n vectors to one
block of size l. Since the run time of ACE depends linearly on the number of
data, clustering with this block selection technique is $(n/l - 1) \times 100\%$ faster
than without. Instead of using a block Y_j we could also reduce the data set
by picking one *segment* Z_j of consecutive data (figure 6.15, left), but this
leads to worse results when the data set consists of points on a trajectory,
as in industrial process data. In this case, each Z_j covers only a part of the
trajectory, while Y_j contains samples from the whole data set. Figure 6.16
(left) shows the two-dimensional projection of a data set from the system
trajectory of a waste paper stock preparation [92]. The middle and right
views show the projections of the segment $Z_1 = \{x_1, x_2, x_3, \ldots, x_{100}\}$ and
of the block $Y_1 = \{x_1, x_{87}, x_{173}, \ldots, x_{8601}\}$, each containing 100 vectors,
respectively. The block Y_1 is a visually better representation for the data
set X than the segment Z_1. Therefore, we restrict our considerations here
to blocks instead of segments or other subsets of X.

   We call the choice of a subset (e.g. a block) of X *preselection*, because
it takes place as a preprocessing step *before* clustering. Preselection reduces the computation time, but it assumes a regular distribution of the
data points, so single outliers (if selected) might cause considerable errors.
Moreover, only a small subset of the original data affects the result at all;
the rest of the data is totally ignored.

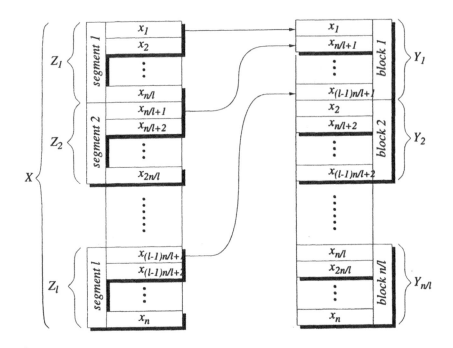

Figure 6.15: Partition of data sets into segments and blocks

To overcome the disadvantages of preselection we *combine* the selection of blocks with the ACE clustering algorithm. Preselection chooses one fixed block from X and uses it to estimate U and V in up to t_{max} repetitive cycles, i.e. in every cycle the same data block is used. Our combined algorithm uses different data blocks in different cycles during cluster estimation, so (almost) all data are considered at least once. This algorithm is shown in figure 6.17. Notice that (except in the first block) the cluster centres obtained from the previous blocks are used as an initialization. The number k_{max} of update steps for each data block can therefore be chosen considerably lower than t_{max}. In our application we even obtained good results with *just one* update of U and V per block ($k_{max} = 1$) and *just one* run through the data set ($i_{max} = 1$). In this case each datum in X is considered exactly once, and we only need n/l updates over all. Generally, partitions U and prototypes V are updated $i_{max} \cdot k_{max} \cdot n/l$ times in contrast to t_{max} times with ACE, so this algorithm is $(t_{max}/i_{max}/k_{max} \cdot l/n - 1) \times 100\%$ faster than ACE. We therefore call this algorithm *fast alternating cluster estimation (FACE)*. FACE can guarantee that each datum is at least considered

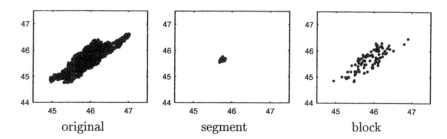

Figure 6.16: Two-dimensional projection of the trajectory data set X (left, 8639 vectors), of the segment Z_1 (middle, 100 vectors), and of the block Y_1 (right, 100 vectors)

once (if FACE is used without preselection). We therefore call FACE a *fair* clustering method. If the FACE update equations for the partitions and the prototypes are chosen according to the fuzzy or possibilistic c-means models, we call this method *fast fuzzy c-means alternating optimization (FFCM-AO)* or *fast possibilistic c-means alternating optimization (FPCM-AO)*, respectively.

initialize prototypes $V^{(0)}$, t=0

for $i = 1, \ldots, i_{max}$

   for $j = 1, \ldots, n/l$

      for $k = 1, \ldots, k_{max}$

      t=t+1

      calculate new partition
      $U^{(t)}(V^{(t-1)}, Y_j, \ldots)$

      calculate new prototypes
      $V^{(t)}(U^{(t)}, Y_j, \ldots)$

Figure 6.17: Fast alternating cluster estimation (FACE)

Notice the difference between preselection and FACE shown in figure 6.18. Preselection (left view) chooses only one fixed block from the data set, which is used for ACE clustering, while FACE (right view) iteratively picks individual blocks, which are considered in only one or a few ACE steps. If the data sets are still too large to be clustered by the FACE algorithm, FACE can be *combined* with preselection. First, a meaningful block is taken from the original data set (preselection) and then this block is clustered using FACE (which updates the clusters using *sub*-blocks then).

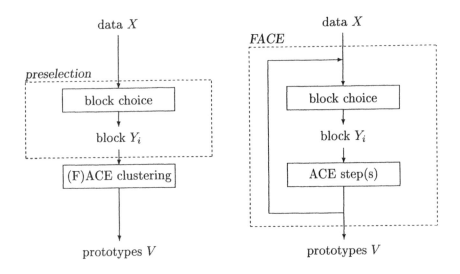

Figure 6.18: Comparison: (F)ACE with preselection (left) and FACE clustering (right)

A modification of FFCM-AO starts with a small sample set $W_0 \subset X$, which is repeatedly increased after a certain number of iterations, so that $W_0 \subset W_1 \subset \ldots \subset W_t \subset X$. This method is called *multistage random sampling fuzzy c-means clustering (mrFCM)* [20, 21]. The same method was applied to the clustering of data sets from municipal energy systems [40]. There, it is referred to as *enhanced fuzzy c-means clustering, EFCM*. This method uses growing sample data sets W_t, which become relatively large after a number of iterations, so that mrFCM/EFCM becomes relatively slow towards the end of the algorithm. In contrast to that algorithm, FACE uses sample blocks that all have the same size, making FACE computationally more efficient than mrFCM/EFCM.

6.6.2   Regular Alternating Cluster Estimation (rACE)

Fuzzy rule based systems can be used to approximate functions $f : \mathbb{R}^p \to \mathbb{R}^q$. These fuzzy systems often use unimodal (e.g. triangular) membership functions that are equidistantly distributed over the input space $\mathbb{R}^p$. These systems are called *regular fuzzy systems* [93]. If the membership functions in the if-part of the rules are specified by the user in advance, we can use *regular fuzzy clustering* [93] to determine the consequent parts of the fuzzy rules from data. The centres of the user-defined membership functions are then interpreted as the input space components of the cluster centres $V \subset \mathbb{R}^p \times \mathbb{R}^q$. The input space components $v_i^{(1)}, \ldots, v_i^{(p)}$, $i = 1, \ldots, c$, of each cluster are left unchanged during clustering, and only the output space components $v_i^{(p+1)}, \ldots, v_i^{(p+q)}$, $i = 1, \ldots, c$, are modified using the conventional prototype functions and the output space components of the data. The prototype function for *regular fuzzy c-means alternating optimization (rFCM-AO)* and *regular possibilistic c-means alternating optimization (rPCM-AO)*, for example, is

$$
v_i^{(j)} = \frac{\sum\limits_{k=1}^{n} u_{ik}^m x_k^{(j)}}{\sum\limits_{k=1}^{n} u_{ik}^m}, \quad i = 1, \ldots, c, \quad j = p+1, \ldots, p+q. \tag{6.37}
$$

ACE instances with arbitrary partition functions and prototype functions modifying only the output space components of the prototypes are called *regular alternating cluster estimation (rACE)* [90]. Notice the lower case "r" for "regular" to distinguish this method from *relational alternating cluster estimation (RACE)* [89]. rFCM-AO, rPCM-AO, and all other instances of rACE use the algorithmic architecture of ACE, but modify only the output space components of the prototypes. rFCM-AO is specified in the ACE model using the prototype function (6.37) and the partition function (6.6). The partition function (6.6) computes the memberships using distances between data and prototypes in the input output product space. In each step of the rFCM-AO loop the output space components of the prototypes are changed by the prototype function and the memberships have to be adapted accordingly. An interesting modification of this algorithm is achieved when the partition function uses only the input space components of data and prototypes:

$$
\mu_i(x) = 1 \Big/ \sum\limits_{j=1}^{c} \left(\frac{\|x^{(1,\ldots,p)} - v_i^{(1,\ldots,p)}\|_A}{\|x^{(1,\ldots,p)} - v_j^{(1,\ldots,p)}\|_A} \right)^{\frac{2}{m-1}}, \quad i = 1, \ldots, c, \tag{6.38}
$$

where $\|.\|_A$ is an appropriate norm. When (6.38) is used as the partition function and (6.37) as the corresponding prototype function, the partition U can be fixed in advance, since the prototypes' input space components $v_i^{(1,\dots,p)}$, $i = 1,\dots,c$, are predefined and fixed by the user. Once having computed the memberships U using (6.38), the *terminal values* of the prototypes can already be computed using the prototype function (6.37). Since (6.37) only affects the *output* space components of V, and (6.38) only uses the *input* space components of V, this clustering algorithm needs only one single ACE loop ($t_{max} = 1$). This and similar regular ACE instances lead to a very fast determination of clusters from data. In the function approximation problem described in [90] the computation time could be reduced by 89% from $125s$ for FCM-AO to $14s$ using a rACE method with (6.37) and pyramid membership functions.

6.7 Comparison: AO and ACE

ACE specifies a very general family of clustering algorithms. AO algorithms are special instances of ACE, whose membership and prototype functions are specified by the first order necessary conditions for extrema of given objective functions. The objective functions seem to lead to a considerable robustness concerning the parameters and the initialization of the cluster centres. For relatively simple objective functions like J_{FCM} (6.1) AO algorithms rapidly find solutions which are close to the global minima. If fuzzy rule based models with a specific membership function shape (triangular, Gaussian) have to be generated, however, a transformation of the membership functions is necessary (see chapter 8). This transformation causes an approximation error which reduces the accuracy of the model. In particular, our experiments have shown that FCM-AO sometimes fails to find the appropriate cluster structures when the data set contains noise and outliers and that PCM-AO runs into problems when the clusters have different point numbers or densities.

"Real" ACE (i.e. non-AO) algorithms are specified by user-defined membership and prototype functions. If the target membership functions are used in the clustering algorithm we can expect more accurate models than with AO algorithms specified by objective functions. Desirable membership function properties like convexity or restriction to local environments can easily be incorporated into ACE. While FCM-AO and PCM-AO are quite sensitive to noise/outliers or different point numbers/densities, respectively, we found that DC, an ACE instance specified by hyperconic membership functions, combines the advantages of both algorithms and leads to the best results in all the examples considered here. However, DC is sensitive to the initialization of the cluster centres, and therefore had to be initialized using some update cycles with FCM-AO. ACE is not restricted to simple prototypes, but can also be used to detect more complex patterns like lines or elliptotypes. For the generation of accurate first order TS function approximators, hyperconic ACE for elliptotypes proved to be superior to the conventional FCE-AO algorithm. When computationally simple membership and prototype functions are used, ACE becomes a very fast and efficient clustering algorithm, making it well-suited for data mining applications dealing with huge data sets. ACE can be further accelerated using block selection (FACE) or regular clustering approaches (rACE).

In general, the robust AO algorithms should be the first choice, when a quick solution has to be found (rapid prototyping). When more accurate models have to be generated, when the data sets contain noise and outliers, or when very large data sets have to be processed in a short time, however, the results can often be improved by non-AO instances of ACE.

Chapter 7

Cluster Validity

Many clustering techniques were developed especially for the recognition of structures in data in higher dimensional spaces. When the data cannot be represented graphically, it is very difficult or sometimes almost impossible for the human observer to determine a partition of the data. But when there is no simple way to decide about a correct or incorrect partition, the following questions and problems have to be considered.

- If all existing clustering algorithms are applied to a data set, one obtains a multitude of different partitions. Which assignment is correct? (see the following example 6)

- If the number of clusters is not known in advance, partitions can be determined for different numbers of clusters. Which partition (or number of clusters) is correct? (see example 7)

- Most algorithms assume a certain structure to determine a partition, without testing if this structure really exists in the data. Does the result of a certain clustering technique justify its application? Is the structure really contained in the data set? (see example 8)

The cluster validity problem is the general question whether the underlying assumptions (cluster shapes, number of clusters, etc.) of a clustering algorithm are satisfied at all for the considered data set. In order to solve this problem, several cluster quality (validity) measures have been proposed. It is impossible to answer all questions without any knowledge about the data. It has to be found out for each problem individually which cluster shape has to be searched for, what distinguishes good clusters from bad ones, and whether there are data without any structure.

185

Figure 7.1: Intuitive partition   Figure 7.2: HCM partition

Example 6   Figure 7.1 shows a data set with 29 data vectors and its intuitive partition into two clusters [31, 10]. (The hard partition is indicated by the empty and solid circles. All horizontally and vertically adjacent data have the distance 1.) Figure 7.2 shows the same data set, however, the left column of the larger cluster was assigned to the smaller one. For $d < 5.5$, the objective function of the HCM algorithm gives now a smaller value than for the partition from figure 7.1. The global minimum leads to an undesired partition in this case. The HCM algorithm tends to *cluster splitting*. (We have already seen in section 2.1 that the partition of clusters of different sizes causes problems for HCM or FCM.)

In order to find out whether an algorithm can be applied to certain kinds of clusters, we have to define test data with which the performance of the technique can be evaluated for characteristic cases. Dunn defines his notion of *good clusters* dependent on a distance function as follows [31]: A classification contains compact, well separated (CWS) clusters if and only if the distance between two data of the convex hull of a class is smaller than the distance between any two data of the convex hull of different classes.

For the example from figure 7.1, a minimum distance can be computed where the two clusters begin to have the CWS property. Dunn introduces a cluster quality measure that is greater than 1 if and only if the partitioned clusters fulfil the CWS condition. The partitions of the HCM which are deficient with respect to this quality measure finally led Dunn to the development of the fuzzified HCM or FCM [31]. Dunn showed with his investigations (without proof) that compact, well separated clusters are significantly better recognized by the FCM. (Strictly speaking, the investigations on the quality of cluster partitions led to the development of the fuzzy c-means.)

If a certain data structure and the corresponding clustering algorithm are chosen, we still have to decide how many clusters we expect in the data set. All algorithms presented so far can only be applied if the number of clusters is known in advance. As already mentioned in section 1.7, cluster validity or quality measures can be used to solve this problem.

Example 7   Continuing example 6, we now use the fuzzy c-means for the detection of compact, well separated clusters. In this special case, we are aware of the fact that we carried out a fuzzification because the fuzzified version of the algorithm provides better results, and not because our data sets cannot be partitioned in a crisp way. A measure for the quality of a partition can thus be the degree of fuzziness inherent in the result. In this case, we consider a partition with crisp memberships as the optimum. (We hope to have assured the geometrically optimal partition by the choice of the algorithm.) If we distribute too many or too few clusters over the data, several clusters have to share a real data cluster or one cluster has to cover several real data clusters. In both cases, this leads to more fuzzy membership degrees. Thus, we run the algorithm for different numbers of clusters, determine the degree of fuzziness of each result and choose the partition that is closest to a hard one. (The quality measure indicated here will be defined and considered in more detail in section 7.1.1.)

Of course, these considerations are only applicable when there are meaningful clusters in the data set.

Example 8   What does a data set without structure look like? A suitable test data set according to Windham [100] consists of data that are uniformly distributed over the unit circle. The requirement for a circular shape results from the Euclidean norm that is used by the fuzzy c-means algorithm. This data set definitely contains no cluster structures. However, the prevailing majority of validity measures yields different values for this test data set when the parameters c and m of the fuzzy c-means algorithm are changed. This means that the validity measures prefer certain parameter combinations to others, although each partition into more than one cluster should be equally bad when there is no cluster structure.

For this data set without structure the validity measure becomes worse and worse for an increasing number of clusters. It is not always clear, however, whether this behaviour reflects the structure or the weakness of the quality function itself. Windham proposes in [100] a computationally complex quality function UDF (uniform data function), which is independent of the parameters c and m of the fuzzy c-means algorithm.

The examples have shown that there are very different kinds of validity measures that are tailored to evaluate specific properties of clustering result. There is no complete theory providing answers to all the mentioned questions. Therefore, most practical procedures are heuristic by nature – to keep the computational complexity feasible. All examples share a definition of an *evaluation function* as a validity measure, which has to be minimized or maximized when searching for a good or correct partition. The main application of validity measures is the determination of the correct number of clusters, as indicated by example 7. Therefore, we examine the properties of various validity measures in the following section.

7.1   Global validity measures

Global validity measures are mappings $g : A(D, R) \to \mathbb{R}$ describing the quality of a complete cluster partition using a single real value. A simple example for such a validity measure is the objective function J (1.7) itself. When we determine the cluster parameters and the memberships, we always try to minimize this objective function. Even if we always find the optimal partition for each number for clusters, the distance between the data and the closest prototypes decreases with an increasing number of clusters. So the objective function J is decreasing, when c is increased. To determine the optimum number of clusters we cannot simply use the minimum or maximum of the objective function. An alternative way to find the optimum number of clusters uses the apex of the objective function (the point with the maximum curvature, called the elbow-criterion in [18]). However, this is a very weak criterion and other functions are used in applications. In the following we introduce a variety of validity measures, discuss their underlying ideas and compare them on the basis of some test examples.

For the main application of global validity measures, the determination of the numbers of clusters, we apply algorithm 3. For this procedure, we assume that the chosen probabilistic algorithm finds the optimal partition for each c so that the optimal partition is also evaluated by the validity measure. This does not hold for any of the introduced algorithms. Consequently, it cannot be guaranteed that the number of clusters determined by this algorithm is correct. (The requirement for a probabilistic clustering algorithm results from the validity measures that usually require properties of probabilistic partitions.)

7.1.1   Solid clustering validity measures

As we saw in example 7, it can be useful to make the partition as crisp as possible. In this case, Bezdek's partition coefficient is a suitable validity measure [10]:

Algorithm 3 (Cluster number determ./global validity measure)

> *Given a data set X, a probabilistic clustering algorithm, and a global validity measure G.*
>
> *Choose the maximum number c_{max} of clusters.*
> $c := 1$, $c_{opt} := 1$, *initialize g_{opt} with the worst quality value.*
> *FOR ($c_{max} - 1$) TIMES*
> $c := c + 1$
>   *Run the probabilistic clustering alg. with the input (X, c).*
>   *IF (qual. measure $G(f)$ of the new partition f is better than g_{opt})*
>   *THEN $g_{opt} := G(f)$, $c_{opt} := c$.*
>   *ENDIF*
> *ENDFOR*
> *The optimum cluster number is c_{opt} and its quality value is g_{opt}.*

Definition 7.1 (Partition coefficient PC) *Let $A(D, R)$ be an analysis space, $X \subset D$, and $f : X \to K \in A_{fuzzy}(D, R)$ a probabilistic cluster partition. The partition coefficient PC of the partition f is defined by*

$$PC(f) = \frac{\sum_{x \in X} \sum_{k \in K} f^2(x)(k)}{|X|}.$$

For any probabilistic cluster partition $f : X \to K$, the inequality $\frac{1}{|K|} \le PC(f) \le 1$ holds because of $f(x)(k) \in [0, 1]$ and $\sum_{k \in K} f(x)(k) = 1$ for $x \in X$ and $k \in K$. In the case of a hard partition, we obtain the maximum value $PC(f) = 1$. If the partition contains no information, i.e. each datum is assigned to each cluster to the same degree, the minimum value is the result. If there is a choice between several partitions, the maximum partition coefficient yields the partition with the "most unambiguous" assignment.

The next validity measure strongly resembles the partition coefficient, however, it is related on Shannon's information theory. The information on the cluster partition consists of the assignment of a datum $x \in X$ to a cluster $k \in K$. The information about the classification is represented by the vector $u = (f(x)(k))_{k \in K}$ for a partition $f : X \to K$. Since we deal with a probabilistic cluster partition, we can interpret the membership u_k as the probability that datum x belongs to cluster k. If we had the information that x definitely belongs to cluster k, then this information is the more valuable the smaller the probability u_k is. Shannon defined the information contents of a single event k, whose probability is u_k as $I_k = -\ln(u_k)$, and the mean information content of a source as the entropy $H = \sum_{k \in K} u_k I_k$.

Definition 7.2 (partition entropy PE) *Let $A(D, R)$ be an analysis space, $X \subset D$, and $f : X \to K \in A_{\mathrm{fuzzy}}(D, R)$ a probabilistic cluster partition. The partition entropy PE of the partition f is defined by*

$$PE(f) = -\frac{\sum_{x \in X} \sum_{k \in K} f(x)(k) \ln(f(x)(k))}{|X|}.$$

In the same way as for the partition coefficient, we can show that $0 \leq PE(f) \leq \ln(|K|)$ holds for any probabilistic cluster partition $f : X \to K$. If f is a crisp partition, we already have the maximum information. The entropy, i.e. the mean information content of a source that tells us the correct partition, is 0. With a uniform distribution of the memberships, however, the entropy increases to its maximum. If we are looking for a good partition, we aim at a partition with a minimum entropy. Bezdek [10] proved the relation $0 \leq 1 - PC(f) \leq PE(f)$ for all probabilistic cluster partitions f. Although this validity measure is based on Shannon's information theory, it is basically a measure for the fuzziness of the cluster partition only, which is very similar to the partition coefficient.

A combinatorial approach via random variables led Windham to another global validity measure. Windham found out that the use of all memberships in the validity measure unnecessarily increases the dependence on the cluster number c. This is the reason why the partition coefficient and the partition entropy tend to increase for larger values of c. He proved that the number of high memberships is drastically reduced with increasing c. His very theoretical approach [99, 10] led to a complex validity measure and only uses the maximum memberships of each datum:

Definition 7.3 (proportion exponent PX) *Let $A(D, R)$ be an analysis space, $X \subset D$, and $f : X \to K \in A_{\mathrm{fuzzy}}(D, R)$ a probabilistic cluster partition with $f(x)(k) \neq 1$ for all $x \in X$ and $k \in K$. The proportion exponent PX of the partition f is defined by*

$$PX(f) = -\ln \left(\prod_{x \in X} \left(\sum_{j=1}^{[\mu_x^{-1}]} (-1)^{j+1} \binom{c}{j} (1 - j\mu_x)^{c-1} \right) \right)$$

where $c = |K|$, $\mu_x = \max_{k \in K} f(x)(k)$ and $[\mu_x^{-1}]$ is the largest integer number smaller or equal to $\frac{1}{\mu_x}$.

Notice that not even a single datum $x \in X$ must be assigned unambiguously to one cluster for the proportion exponent. Then we would have

$\mu_x = 1$ and $[\mu_x^{-1}] = 1$, so that the sum for this x would yield zero and thus the whole product would be zero. But then the logarithm is not defined. Although we aim at hard partitions, we must not actually reach them, as long as we use this validity measure. In order to be able to use the proportion exponent at all in that paradoxical situation, we will use a higher fuzzifier m. (Remember that the closer m is to 1, the harder the resulting partition is.)

Neglecting the logarithm, the complicated expression PX can be interpreted as a measure for the number of partitions that classify *all* data vectors *better* than the considered partition. A better classification of an already unambiguously assigned datum is of course impossible. This explains why the proportion exponent does not admit crisp assignments. However, it also means that the validity measure PX already indicates an optimum partition, when just one single datum is assigned unambiguously. (On the other hand, the validity measures PC and PE indicate an optimum partition only if *all* data are assigned unambiguously.) The negative logarithm means that we are looking for a maximum value for the optimum partition.

If we use a higher fuzzifier (e.g. $m = 4$), crisp assignments are almost impossible in practice. Notice, however, that we multiply n numbers smaller than one, where n is the number of elements of the data set X. Thus we usually obtain values in the order of 10^{-200}, and therefore have to use an implementation supporting large exponents.

Let us once again come back to the notion of CWS clusters according to Dunn from example 6. In order to avoid complications because of the notion of the convex hull, we only consider CS clusters: A partition contains compact, separated (CS) clusters if and only if any two data from one class have a smaller distance than any two data from different classes. We require for CS clusters that their distance from each other is larger than their maximum diameters.

Definition 7.4 (Separation index D_1) *Let $A(D, R)$ be an analysis space, $X \subset D$, $f : X \to K \in A(D, R)$ a hard cluster partition, $A_k := f^{-1}(k)$ for $k \in K$, $d : D \times D \to \mathbb{R}_+$ a distance function. The separation index D_1 of the partition f is defined by*

$$D_1(f) = \min_{i \in K} \left\{ \min_{j \in K \setminus \{i\}} \left\{ \frac{d(A_i, A_j)}{\max_{k \in K} diam(A_k)} \right\} \right\},$$

where $diam(A_k) = \max\{d(x_i, x_j) \mid x_i, x_j \in A_k\}$, and the distance function d is extended to sets by $d(A_i, A_j) = \min\{d(x_i, x_j) \mid x_i \in A_i, x_j \in A_j\}$ (for $i, j, k \in K$).

If the separation index of a given hard partition is greater than one, we thus deal with a partition in CS clusters. Larger values of the separation index indicate good partitions. In order to be able to use this validity measure for our algorithms, too, which do not provide a hard partition, the induced hard partition could be used for the determination of the separation index, partly neglecting information inherent in the fuzzy partition for the validity measure.

For large data sets, the determination of the separation indices is computationally very expensive because the number of operations for the determination of the diameters and distances of the clusters quadratically depends on the number of data.

In contrast to the validity measures presented before, the separation index takes the data themselves into account, not only the membership degrees. A compensatory validity measure could take both properties – the crispness of a partition and the geometrical structure of the clusters and data – into account. However, the separation index can only evaluate hard partitions and is therefore not suited as a compensatory validity measure. Xie and Beni [101] propose a compensatory validity measure that is based on the objective function J by determining the average number of data and the square of the minimum distances of the cluster centres. Here, of course, the existence of cluster centres is needed. The clustering algorithms introduced here provide cluster centres, but this is not necessarily the case with other clustering methods.

Definition 7.5 (Separation S) *Let $A(D,R)$ be an analysis space, $X \subset D$, $f : X \to K \in A_{\text{fuzzy}}(D,R)$ a probabilistic cluster partition, and d a distance function. The separation S of the partition f is defined by*

$$S(f) = \frac{\sum_{x \in X} \sum_{k \in K} f^2(x)(k) d^2(x,k)}{|K| \min\{d^2(k,l) \mid k,l \in K, k \neq l\}}.$$

When applying the fuzzy c-means algorithm with a fuzzifier $m \neq 2$, m can alternatively be inserted into the exponent for the membership. Xie and Beni proved the inequality $S \leq \frac{1}{D_1^2}$ for the corresponding hard partition, establishing a relation to a partition into CS clusters according to Dunn. Compared to D_1, the validity measure S has the advantage of a much simpler computation and it takes the membership degrees into account so that a detour via a hard partition is unnecessary.

If we have further requirements with respect to the clustering techniques, e.g. the description of cluster shapes by covariance matrices (cf. sections 2.2 and 2.3), further individual validity measures can be defined. Gath and Geva propose three such measures together with their clustering algorithm in [37].

Definition 7.6 (Fuzzy hypervolume FHV) *Let $A(D, R)$ be an analysis space, $X \subset D$, $f : X \to K \in A_{\text{fuzzy}}(D, R)$ a probabilistic cluster partition by the algorithm of Gath and Geva. Using the notations of remark 2.3, we define the fuzzy hypervolume FHV of the partition f by*

$$FHV(f) = \sum_{i=1}^{c} \sqrt{\det(A_i)}.$$

Definition 7.7 (Average partition density APD) *Let $A(D, R)$ be an analysis space, $X \subset D$, $f : X \to K \in A_{\text{fuzzy}}(D, R)$ a probabilistic cluster partition by the algorithm of Gath and Geva. Using the notations of remark 2.3, we define the average partition density APD of the partition f by*

$$APD(f) = \frac{1}{c} \sum_{i=1}^{c} \frac{S_i}{\sqrt{\det(A_i)}},$$

where $S_i = \sum_{j \in Y_i} f(x_j)(k_i)$ and

$$Y_i = \{j \in \mathbb{N}_{\leq n} \mid (x_j - v_i)^{\top} A_i^{-1} (x_j - v_i) < 1\}.$$

Definition 7.8 (Partition density PD)
Let $A(D, R)$ be an analysis space, $X \subset D$, and $f : X \to K \in A_{\text{fuzzy}}(D, R)$ a probabilistic cluster partition by the algorithm of Gath and Geva. Using the notations of remark 2.3 and of the previous definition, we define the partition density PD of the partition f by ·

$$PD(f) = \frac{\sum_{i=1}^{c} S_i}{FHV(f)}.$$

The sum of all cluster sizes is considered by the fuzzy hypervolume. Since all data vectors are covered by the clusters in any case, a minimum of this validity measure indicates small, compact clusters, which just enclose the data. The partition density corresponds to the physical notion of density (number of data per volume), and it has to be maximized because the clusters should correspond to distinct point accumulations. The same is valid for the average partition density, which is determined as the mean value of the (physical) densities in the cluster centre. The number of data per cluster is considered by the terms S_i for a cluster $k_i \in K$. For this purpose the memberships of the data are added up that have a distance from the cluster centre smaller than one with respect to the A_i^{-1} norm. These are the data vectors whose distance to the centre does not exceed

the average distance of the assigned data. Hence, we implicitly require that
most data should be near the prototype (so that the sum contains many
terms) and that these data should be unambiguous (in order to add high
memberships). A good cluster partition in the sense of these three validity
measures thus consists of well separated clusters with a minimum overall
volume and an accumulation of data vectors within the cluster centres.

The suitability of the validity measures introduced will be tested now
by some examples. The procedure for the determination of the validity
measures presented was partly different. PC, PE, and S can be determined
after an FCM run with $m = 2$, however, PX is undefined in most cases. For
FHV, APD, and PD, a GG run has to be performed, leading to almost
hard partitions and almost constant validity measures PC and PE, and
an undefined PX. For this reason, PC, PE, and $S1 = S$ were determined
from an FCM run with $m = 2$, PX from an FCM run with $m = 4$, and
FHV, APD, PD, and $S2 = S$ from a GG run with $m = 2$. In addition, the
tested data sets themselves are chosen so that the fuzzy c-means algorithm
can find a suitable partition: we did not use different shapes and sizes of the
clusters because we do not want to compare the abilities of the algorithms,
but properties of the validity measures.

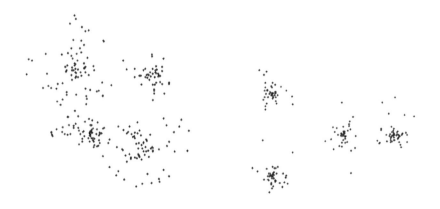

Figure 7.3: Test pattern A   Figure 7.4: Test pattern B

The data set already known from figure 2.1 was used as a test pattern
(figure 7.3). In addition, figures 7.4, 7.5, and 7.6 show three similar test
data sets where the three clusters on the left-hand side get closer and closer
to each other. Finally, they are so close in figure 7.6 that the fuzzy c-means
algorithm with $m = 2$ and $c = 4$ does not divide the point cloud on the left-
hand side into three clusters, but covers the left and right sides of the test

Figure 7.5: Test pattern C   Figure 7.6: Test pattern D

data with two clusters each. An optimum quality value for $c = 4$ cannot be expected for such a case. For the algorithm of Gath and Geva, the partition of the data from figure 7.4 causes problems for $c > 5$, since the clusters were already very well and unambiguously separated with $c = 4$ and all following clusters cover only some single outlier data. Here, the cluster sizes become so small that there can be numerical overflows. If clusters are defined by an individual (extreme) outlier, we can obtain (almost) singular covariance matrices. If the algorithm terminates with such an error, we assume that a search for a greater c makes no sense. This is the case for $c > 5$, hence the quality values FHV, APD, PD, and $S2$ were continued with a constant value.

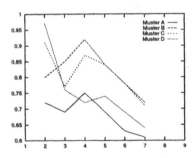

Figure 7.7: PC (maximize)

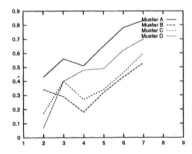

Figure 7.8: PE (minimize)

When we compare the validity measures, the partition coefficient PC (figure 7.7) and the partition entropy PE (figure 7.8) behave in a very similar manner, just as expected. If we assume the optimum cluster number to be $c = 4$ in all examples, both validity measures recognize the correct number only for the well separated data set from figure 7.4. The partition coefficient also has its global maximum for the data from figure 7.3 and $c = 4$. For the data set from figure 7.5, both validity measures at least still indicate a local extremum. However, they prefer the less ambiguous

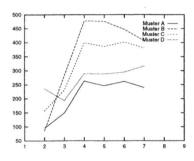

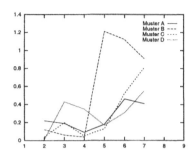

Figure 7.9: PX (maximize)   Figure 7.10: $S1$ (minimize)

partition into two clusters. The monotonicity with increasing c can be clearly observed for both validity measures. (When comparing the validity measures, the different evaluation criteria have to be taken into account – the partition coefficient has to be maximized, while the partition entropy has to be minimized.) If a comparison of the two validity measures is made with respect to their behaviour for the data from figure 7.3, the partition coefficient seems to provide better results, since a partition into four clusters is intuitively preferred to a partition into two clusters.

The proportion exponent PX (figure 7.9) does not suffer from the monotonicity with increasing c, but indicates the correct cluster number for the data from figures 7.3 and 7.4 only. In the other two cases, the validity measure PX starts with rather low values, jumps to a local maximum for $c = 4$, and remains approximately at this higher lever whereby the global maximum is reached once for $c = 6$ and once again for $c = 7$ (with respect to $c \in \mathbb{N}_{\leq 7}$). Notice that the differences of the validity measures are low for $c > 4$, so partitions with $c > 4$ clusters are interpreted as good partitions, too, but with more clusters than necessary. In this sense, PX points to an optimum number $c = 4$ for these data sets as well. For the data from figure 7.6, this validity measure takes advantage of the fact that it was determined with $m = 4$. Then, the fuzzy c-means algorithm divides the left point cloud into three clusters, which was not the case for $m = 2$.

The separation S ($S1$ for FCM in figure 7.10, $S2$ for GG in figure 7.11) clearly shows extreme values with $c = 4$ for data sets that are easy to cluster. For the data set from figure 7.5, the local minima for $c = 4$ are only a little greater than the global minima for $c = 2$. This validity measure only slightly prefers the partition into two clusters to the partition into four clusters. This behaviour is the same – though weakened – for S2 and the data from figure 7.6. If we ignore the results of $S1$ for the data from 7.6 (FCM with $m = 2$ provided a bad partition for $c = 4$), the separation S

seems to be quite practical for the detection of the cluster number in the case of well-separated clusters.

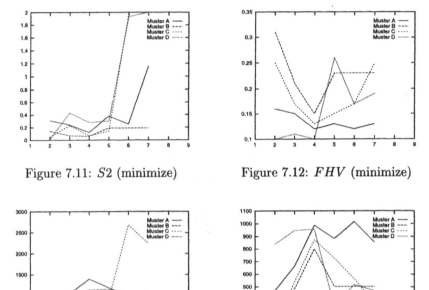

Figure 7.11: $S2$ (minimize)   Figure 7.12: FHV (minimize)

Figure 7.13: APD (maximize)   Figure 7.14: PD (maximize)

Finally, let us consider the triple of validity measures by Gath and Geva. One validity measure for each test data set did not indicate an extreme value for the optimum number $c = 4$. For the data from figures 7.3 and 7.6, respectively, the fuzzy hypervolume had the same value for $c = 4$ and $c = 6$, and for $c = 2$ and $c = 4$, respectively, with a precision of $\frac{1}{100}$. Thus, two validity measures for each of the four test data sets had their (global) extremum for $c = 4$. (The GG algorithm was initialized with the result from FCM with $m = 4$ for the data from figure 7.6 in order to avoid the bad initial partition with $m = 2$.) As already mentioned, problems can occur when – after a good partition has been found – further clusters cover only a little noise data. These clusters have a very small volume and a high density. The high density can have a strong impact on the average value APD of the densities. Figure 7.13 shows the results for $c > 5$ for the data from figure 7.5. This does not affect the overall partition density PD, which has the global maximum for $c = 4$ for all three data sets from figures 7.4, 7.5, and 7.6. For the data set from figure 7.3, the global maximum was just missed by the value for $c = 6$, although there can certainly be a dispute about the *correct* partition. From these three validity measures, PD is distinguished, but the best results can be obtained using all measures in combination (choose the number with which most validity measures indicate extreme values).

It is obvious that the validity measures which also consider parameters of the cluster prototypes provide better results. There is no *best* validity measure as can be seen in figure 7.6: depending on the application, two unambiguous or four fuzzy clusters can be seen here. Depending on the desired results, a validity measure should be chosen for the respective application.

If the algorithms by Gustafson-Kessel or Gath-Geva are used for the detection of straight line segments, the validity measures presented above can also be used for linear clustering algorithms.

7.1.2 Shell clustering validity measures

In principle, the considerations from the previous section also apply to shell clusters. Because of the cluster overlaps we do not obtain hard memberships near the intersection points, even with an optimum partition, but most of the data should be classified unambiguously. Because of the flexible contours of the shell clusters, the data can be covered well without really recognizing one single cluster; an example for this problem is shown in figure 4.31. Hence, special validity measures are needed for shell clusters. The above considered validity measures yielded the best results, when we included as much information as possible about the cluster prototypes into the validity measure. However, the validity measures introduced so far can not be easily adopted here since the prototypes have different meanings. For the compact clusters of the algorithm by Gath and Geva, a good cluster should have a volume as small as possible, hence the determinant of the fuzzy covariance matrix is minimized. For some shell clustering algorithms, there also occur positive definite matrices as parameters of the prototypes. However, these implicitly describe the orientation and radii of the recognized ellipses. The decision as to whether it is a well or badly recognized ellipse should be independent of the ellipses' radii. A minimization of the determinant makes no sense here.

Davé [24] defined the Gath-and-Geva validity measures hypervolume and partition density for the case of circle and ellipse clusters. Krishnapuram, Frigui, and Nasraoui generalized these validity measures to arbitrary shell clusters [66]: the distance between the data and the cluster is essential for the quality of a cluster. While it can be easily determined for compact clusters by the distance between the datum and the position vector of the prototype, the determination of the distance is more complicated for shell clusters. We need the closest point z on the cluster contour for an arbitrary point x. Their Euclidean distance is the distance between datum x and the shell cluster. In the special case of a circle cluster with centre v and radius r, we have $z = v + r\frac{x-v}{||x-v||}$. For other shapes, the determination of the point z can be very difficult and expensive. We approximately determined the distance $||x - z||$ for the FCES algorithm (section 4.4) and the point z for the MFCQS algorithm (section 4.7).

For a probabilistic cluster partition $f : X \to K$, we now define the vector $\Delta_{i,j} = x_j - z_{i,j}$, for $x_j \in X$ and $k_i \in K$, where $z_{i,j}$ denotes the point on the contour of the cluster k_i that is closest to x_j. The vectors $\Delta_{i,j}$ represent the direction and size of the deviation of datum x_j from cluster k_i. If the cluster is a point, they correspond to the vectors $x_j - v_i$ that gave the dispersion of the datum around the cluster centre v_i for compact clusters and from which we formed the covariance matrices. Thus, we analogously define:

Definition 7.9 (Fuzzy shell covariance matrix) *Let $A(D, R)$ be an analysis space, $X \subset D$, $f : X \to K \in A_{\text{fuzzy}}(D, R)$ a probabilistic cluster partition and $\Delta_{i,j}$ the distance vector for $x_j \in X = \{x_1, x_2, \ldots, x_n\}$ and $k_i \in K = \{k_1, k_2, \ldots, k_c\}$ of the datum x_j from cluster k_i. Then,*

$$A_i = \frac{\sum_{j=1}^{n} f^m(x_j)(k_i) \; \Delta_{i,j}\Delta_{i,j}^\top}{\sum_{j=1}^{n} f^m(x_j)(k_i)} \tag{7.1}$$

is called the fuzzy shell covariance matrix of cluster k_i.

This matrix is a measure for the dispersion of the data around the cluster contour. So the volume and density validity measures by Gath and Geva are now applicable also to shell clusters. (Instead of the covariance matrix from remark 2.3, we use the matrix (7.1) in the definitions 7.6 and 7.7.)

In addition, the distance $||\Delta_{i,j}||$ between a datum and a cluster contour can be interpreted as the thickness of the contour:

Definition 7.10 (Shell thickness T) *Let $A(D, R)$ be an analysis space, $X \subset D$, $f : X \to K \in A_{\text{fuzzy}}(D, R)$ a probabilistic cluster partition and $\Delta_{i,j}$ be the distance vector for $x_j \in X$ and $k_i \in K$ of datum x_j from cluster k_i. We define T_S as the sum of the mean shell thicknesses T_S of the partition f:*

$$T_S(f) = \sum_{i=1}^{c} \frac{\sum_{j=1}^{n} f^m(x_j)(k_i) \; ||\Delta_{i,j}||^2}{\sum_{j=1}^{n} f^m(x_j)(k_i)}.$$

If the clusters $k_i \in K$ are circle clusters with centres v_i and radii r_i, we define the mean circle thickness T by

$$T_C(f) = \frac{T_S(f)}{\frac{1}{|K|}\sum_{i=1}^{c} r_i}$$

where $\|\Delta_{i,j}\|^2 = (\|x_j - v_i\| - r_i)^2$.

For arbitrary shell clusters, T_S is a suitable validity measure, even if a fluctuation $\|\Delta_{i,j}\| = \frac{1}{10}$ with a shell diameter 1 has certainly more importance than with a shell diameter 10. If information about the shell size is available, such as for circle clusters, this should be included like in the average circle thickness T_C.

As mentioned in the beginning, the correctness of the cluster number determined by global validity measures essentially depends on the quality of the cluster partition. The considerations on shell clustering in section 4 showed that even with a correct cluster number, only moderate results can be expected without a good initialization (cf. figures 3.4, 4.5, 4.26, and 5.16 as examples). If the partition with the correct cluster number is not clearly distinguished from the other partitions by its quality, we cannot expect the validity measures to clearly distinguish the optimum number of clusters. The unsatisfying results of the global shell clustering validity measures motivated the development of alternative techniques that are introduced in the following section.

7.2   Local validity measures

The determination of the optimum cluster number using global validity measures is very expensive, since clustering has to be carried out for a variety of possible cluster numbers. Here, especially for shell clustering, it can be that just the analysis with the correct cluster number does not recognize one single cluster correctly. If the perfect partition cannot be recognized in a single run, we could at least filter the presumably correctly recognized clusters. Here, a validity measure is necessary that evaluates single clusters, a so-called local validity measure. Subsequently, we can investigate the badly recognized data separately. If we use the memberships for the detection of good clusters, we have to consider the case that several prototypes share one cluster. In a probabilistic cluster partition, none of the data vectors obtain memberships clearly greater than $\frac{1}{2}$ such that a local validity measure would incorrectly conclude that there were two badly recognized clusters. (An example for a case like this is shown in figure 4.6; there, the upper left semi-circle is covered by two very similar clusters.)

While in the previous sections we tried to examine the space of possible cluster numbers by global validity measures completely, we now try to approach the correct partition more directly. The idea behind local validity measures is the following: by analysing the final partition of the previous run we provide improved initializations for the next run. We continue this procedure until finally the whole partition consists of good clusters only. When the final partition consists only of good clusters, we have found the optimum number of clusters, too.

An algorithm whose strategy is very similar, although it has basically nothing to do with local validity measures, is the CCM algorithm for the detection of straight line segments described in the following and based on the algorithm by Gustafson and Kessel. We consider it here since it works well for the determination of the number of straight line segments and the subsequent algorithms have similar underlying ideas.

7.2.1 The compatible cluster merging algorithm

The compatible cluster merging algorithm (CCM) by Krishnapuram and Freg [61] identifies the number of straight lines in a two-dimensional data set (or planes in the three-dimensional case). As we have seen in section 3.3, the algorithms by Gustafson and Kessel or Gath and Geva detect straight lines as well as special algorithms for line recognition. Hence, the GK algorithm serves as a basis of the CCM algorithm.

Just as with the technique for global validity measures, an upper limit c_{max} of the cluster number has to be chosen. This limit must not be a good estimation for the actual cluster number but be clearly above it. On the other hand very large values for c_{max} increase the computing time of the algorithm considerably. The undesirable covering of collinear straight line segments by a single cluster is aimed to be avoided by the high cluster number since the data space is partitioned into smaller clusters by the high number c_{max}. For the same reason, however, the opposite case can occur, where a single long straight line is covered by several clusters (example in figure 7.15). In cases like this, we will refer to them as *compatible* clusters and merge them into one cluster. Thus, a probabilistic Gustafson-Kessel run with c_{max} clusters is done first. The resulting clusters are compared to each other in order to find *compatible* clusters. This way, the number of clusters is reduced correspondingly. We form groups of compatible clusters, merge them and determine new Gustafson-Kessel prototypes for the merged groups. With the new, often clearly decreased cluster number and the merged clusters as an intitialization, we run the GK algorithm once more, form and merge compatible clusters, until finally all clusters are incompatible and no clusters can be merged any more. We consider the cluster partition determined this way as the optimum partition. Here, the computing time decreases with each further GK run because the clusters get fewer and fewer, and we always have a very good initialization. In addition, GK runs are needed for very few cluster numbers compared to algorithm 3. This technique leads to a better result in a shorter time, as we will see later on.

When do we call two clusters of the Gustafson-Kessel algorithm compatible? On the one hand, two compatible clusters should form the same hyperplane (i.e. lie on a straight line in $\mathbb{R}^2$), and on the other hand, they should be *close* to each other in relation to their size.

Definition 7.11 (Compatibility relation on GK clusters) *Let* $p \in \mathbb{N}$ *and* $C := \mathbb{R}^p \times \{A \in \mathbb{R}^{p \times p} \mid \det(A) = 1,\ A \ symmetric \ and \ positive \ definite\ \}$. *For each* $k_i = (v_i, A_i) \in C$, *let* λ_i *be the largest eigenvalue of the matrix* A_i *and* e_i *the normalized eigenvector of* A_i *associated with the smallest eigenvalue. For each* $\gamma = (\gamma_1, \gamma_2, \gamma_3) \in \mathbb{R}^3$ *a CCM compatibility relation* $\doteq_\gamma \subseteq C \times C$ *is defined by*

$$\forall k_1, k_2 \in C : \quad k_1 \doteq_\gamma k_2 \quad \Leftrightarrow \quad |e_1 e_2^\top| \geq \gamma_1 \quad \wedge \tag{7.2}$$

$$\left| \frac{e_1 + e_2}{2} \frac{(v_1 - v_2)^\top}{||v_1 - v_2||} \right| \leq \gamma_2 \quad \wedge \tag{7.3}$$

$$||v_1 - v_2|| \leq \gamma_3 \left(\sqrt{\lambda_1} + \sqrt{\lambda_2} \right). \tag{7.4}$$

Two clusters k_1, $k_2 \in C$ form (approximately) the same hyperplane if the normal vectors of the two hyperplanes are parallel and each origin is in the other hyperplane. The normal vector of the hyperplane described by a cluster k_1 corresponds to the eigenvector e_1 from the definition. If e_1 and e_2 are parallel, their scalar product is one. When we choose $\gamma_1 = 1$, equation (7.2) requires the hyperplanes to be parallel. If the two normal vectors are (almost) parallel, $\frac{1}{2}(e_1 + e_2)$ is approximately the mean normal vector of the two hyperplanes. When k_1 and k_2 form the same hyperplane, the vector that leads from the origin of one plane to the origin of the other one must lie within the common hyperplane. This is the case when the difference vector is orthogonal to the normal, i.e. the scalar product of the two vectors is zero. If we choose $\gamma_2 = 0$, equation (7.3) corresponds to this requirement. If (7.2) and (7.3) are valid with $\gamma_1 = 1$ and $\gamma_2 = 0$, the GK clusters k_1 and k_2 span the same hyperplane.

We merge two clusters just in case there is no *gap* between them so that they are adjacent to each other. Therefore, let us consider the data as realizations of two uniformly distributed random variables Z_1 and Z_2. In the projection onto the common straight line, the cluster position vector then represents the assumed mean. The variance σ_i^2 of the random variables Z_i, $i \in \{1, 2\}$ is the largest eigenvalue of the covariance matrix of the corresponding cluster. In general, the relation $\sigma^2 = \frac{L^2}{12}$ holds for uniform distributions over an interval with length L. The interval thus stretches $\frac{L}{2} = \sqrt{3}\,\sigma$ in both directions from the assumed mean. If the distance of the assumed means of both random variables is less than the sum of both half interval lengths, the random variables or straight line segments overlap. In this case, we will consider the two clusters as belonging together and merge them. For $\gamma_3 = \sqrt{3}$, that corresponds to condition (7.4).

Because of the noise in the data along the straight lines or planes, the constants γ_1 to γ_3 should be weakened in order to be able to identify mergable clusters at all. In [61] Krishnapuram and Freg suggest $\gamma_1 = 0.95$, $\gamma_2 = 0.05$, and $\gamma_3 \in [2, 4]$.

Note that the relation just defined is not an equivalence relation. This is caused by the use of inequalities in the definition. If we distribute three linear clusters of the same size on one straight line equidistantly, adjacent clusters may be in a relation to each other (depending on γ_3); however, this does not apply to the two outer clusters. The relation $\doteq_\gamma$ is hence not always transitive. That means that there is not always an unambiguous factorization of the linear clusters with respect to that relation. However, this circumstance has no influence on the practical use of the heuristic CCM algorithm. For the implementation, it should be assured that all clusters of the same group are pairwise compatible. It is not sufficient that a cluster is in relation with one element of a group only, in order to be merged with this group. Which group combination of the remaining ones is finally chosen, however, does not play a crucial role for the final result of the CCM algorithm.

Furthermore, a special case has to be considered for the implementation of condition (7.3): if the two eigenvectors are nearly identical except for their signs, they eliminate each other when they are added. Depending on the applied technique for the determination of the eigenvectors, two eigenvectors for two parallel straight lines can be different with respect to their signs only. Then conditions (7.3) and (7.2) hold because of the parallelity and the elimination, respectively. Also, condition (7.4) can now hold even when the clusters are not in the same hyperplane. This case has to be treated separately: if the sum of the eigenvectors in the Euclidean norm is significantly smaller than one, an elimination occurred and the vectors should be subtracted instead of being added.

If the algorithm is applied to the data sets from chapter 3 with $c_{\max} = 12$ and $\gamma = (0.95, 0.05, 2.0)$, the results are mainly correct. However, sometimes two compatible clusters are not merged: when two clusters describe

Algorithm 4 (CCM algorithm)

Given a data set X and a compatibility relation $\doteq$ on GK clusters.

Choose the maximum number $c_{\max}$ of clusters.

$c_{\text{new}} := c_{\max}$
REPEAT
      $c_{\text{old}} := c_{\text{new}}$
      Run the GK algorithm with (X, c_{new}).
      Determine eigenvectors/eigenvalues of the covariance matrices.
      Form (transitive) groups of clusters with respect to the relation $\doteq$.
      FOR (each group with more than one cluster)
            Merge all clusters of the group into one single cluster.
            Compute the position and orientation of the new cluster.
            Update the number of clusters in c_{new}.
      ENDFOR
UNTIL $(c_{\text{old}} = c_{\text{new}})$

the same straight line segment and the prototypes are close together. This happens when the data are scattered along a straight line, but have a large variance. The actual straight line segment is not approximated by two short clusters in a row (figure 7.15) but two long parallel ones (figure 7.16). Here, the position vectors of two clusters k_1 and k_2 move closer together, and the difference vector of the positions $\Delta v = v_1 - v_2$ is (approximately) perpendicular to the direction vectors. For this reason, condition (7.3) prevents a merging of the clusters. (Figures 7.15 and 7.16 show the extreme positions only; condition (7.3) does not hold for the intermediate positions either.)

Figure 7.15: Collinear clusters   Figure 7.16: Parallel clusters

This problem is avoided by the following compatibility relation between GK clusters. Instead of the condition (7.3) ensuring the perpendicularity proposed by Gath and Geva, we directly measure the distance between the prototypes in the direction of the mean normal. If this is smaller than a certain maximum distance, we assume compatible clusters. If the mean distance δ from (5.15) is used as the maximum distance, only those parallel clusters are merged that have a smaller distance than the data among themselves.

Definition 7.12 (modified compatibility relation on GK clusters) *Let* $p \in \mathbb{N}$ *and* $C := \mathbb{R}^p \times \{A \in \mathbb{R}^{p \times p} \mid \det(A) = 1,$ A *symmetric and positive definite}. For each* $k_i = (v_i, A_i) \in C$, *let* λ_i *be the largest eigenvalue of matrix* A_i *and* e_i *the normalized eigenvector of* A_i *associated with the smallest eigenvalue. For each* $\gamma := (\gamma_1, \gamma_2, \gamma_3) \in \mathbb{R}^3$ *a modified CCM compatibility relation* $\doteq_\gamma \subseteq C \times C$ *is defined by*

$$\forall k_1, k_2 \in C : \quad k_1 \doteq_\gamma k_2 \quad \Leftrightarrow \quad |e_1 e_2^\top| \geq \gamma_1 \quad \wedge$$
$$\left| \frac{e_1 + e_2}{||e_1 + e_2||} (v_1 - v_2)^\top \right| \leq \gamma_2 \quad \wedge \quad (7.5)$$
$$||v_1 - v_2|| \leq \gamma_3 \left(\sqrt{\lambda_1} + \sqrt{\lambda_2} \right).$$

Figure 7.17: Floppy disk box

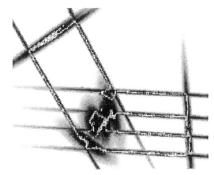

Figure 7.18: CCM analysis

With the modified compatibility relation $\doteq_{(0.95, \delta, 2.0)}$, the correct cluster number is obtained for the data sets from figures 3.1 and 3.9. This relation, however, is sensitive to additional noise data, because GK clusters approximate these with compact clusters. The compatibility relation cannot handle this case: point clouds cannot be merged. Noise data are mostly approximated by numerous GK clusters by the CCM algorithm. Without noise clouds, the results of the modified CCM algorithm are mostly very good.

Figure 7.19: Matchbox   Figure 7.20: CCM analysis

Figures 7.17, 7.19, and 7.21 show grey value images from which the
data sets in figures 7.18, 7.20, and 7.22 were generated by contrast intensi-
fication, a contour operator (Sobel) and threshold techniques. If there are
unambiguous straight lines in the data sets, these are well recognized. In
the areas of many small straight lines (middle view in figure 7.18) the *more
effective* ellipsoidal shape is used instead of an approximation by GK clusters in straight line shapes. In order to make this undesired classification
more accurate, data that are partitioned by such ellipsoidal clusters can be
extracted and partitioned once more in a separate GK run. Furthermore,
in some cases the extension of the clusters into the direction of the straight
line prevents the detection of other straight lines. For several straight lines
that run parallel and closely together in particular, perpendicular straight
lines are also covered (cf. figure 7.20). An improvement of the recognition
performance can be achieved with methods from section 7.3 in this case.

An example for a very good performance is shown in figures 7.21 and
7.22. The CCM algorithm was started with $c_{max} = 40$, 28 clusters were
recognized. Even smaller straight line segments were covered by single
clusters in this example. The only weak point is the massive upper contour
of the hole in the middle of the puncher's lever. Since the volume of the GK
clusters is limited, the thick straight line could not be covered by a single
GK cluster. The three single clusters do not follow the thick straight line
now, but are nearly horizontal such that the compatibility relation does
not initiate a merging, because of the large distance between the straight
lines from each other. If point accumulations like this cannot be avoided,
an algorithm for line thinning [60] should be applied in order to prevent
this effect.

Clusters that cover extremely thick contours as in figure 7.22 or many
small straight line segments as in figure 7.18 can be recognized using eigenvalues of the GK covariance matrices. The relation of the large eigenvalue

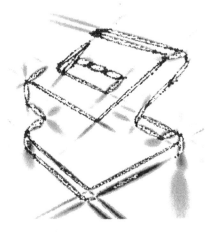

Figure 7.21: Puncher   Figure 7.22: CCM analysis

to the small one is 1 for circular shaped clusters; the larger it is, the more the cluster is stretched. If the value falls below a minimum for this relation, it can be concluded that there is a compact cluster.

7.2.2 The unsupervised FCSS algorithm

The CCM algorithm did not evaluate the clusters; a local validity measure was unnecessary. If a data set contained only straight line segments, each GK cluster approximated (part of) a line segment. Shell clustering algorithms (we start with considering circles) do not behave that nicely. There are many more minimum partitions than for clustering straight lines. Not every circle cluster represents a correctly recognized circle segment, but instead might cover data vectors from different circles (cf. figures 4.4 or 4.5). Simply merging several circles does not give any advantage here, because the union contains not only the data of one but of several circles.

With the unsupervised fuzzy c-spherical shells algorithm (UFCSS) by Krishnapuram, Nasraoui, and Frigui [67], further steps are therefore carried out in addition to merging clusters. After the FCSS algorithm is completed with a maximum cluster number c_{max}, groups of compatible clusters are first searched for and merged once again. Afterwards, especially good clusters are identified using a local validity measure, and all data that represent these clusters are eliminated from the data set under consideration. Thus, the number of data and clusters for the following iterations are decreased, and it is simpler for the FCSS algorithm to find the correct position of the remaining clusters. Finally, those clusters covering only very few data vec-

tors are deleted without substitution. This procedure is repeated until none
of these operations can be applied any longer. Then, the good clusters, together with the removed data vectors, are considered again for fine-tuning,
and the merging of compatible and the deletion of small clusters is iterated
anew, until again no changes occur. When the algorithm is terminated, we
have the (hopefully) optimum partition of the data set.

As a local quality function we use the data density near the shell, a
validity measure that resembles the mean partition density for GK clusters,
without determining the mean over the clusters. Instead of the covariance
matrices of the GK algorithm, we use shell covariance matrices according to
(7.1), which determine the scattering of the data around the cluster contour.
A shell covariance matrix A induces a norm as in the GK algorithm. This
norm determines whether a certain data vector is more distant from the
contour than the average of the data assigned to the cluster or not. For a
test datum x with a distance vector Δ from the cluster contour, we have
$\Delta^\top A^{-1}\Delta < 1$, if x is closer to the contour than the average of the data
vectors. The sum of the memberships of all data that fulfil this condition is
a measure for the quality with which the data *match* the cluster contour or
how well the contour approximates the data. If we again consider $\sqrt{\det(A)}$
as the volume of the cluster contour, we obtain the shell density as a local
validity measure:

Definition 7.13 (Shell density SD) *Let $A(D, R)$ be an analysis space,*
$X \subset D$, $f : X \to K \in A_{\text{fuzzy}}(D, R)$ a probabilistic cluster partition and $\Delta_{i,j}$
for $x_j \in X$ and $k_i \in K$ the distance vector between datum x_j and cluster
k_i. We define the shell density SD_i of cluster k_i in the partition f by

$$SD_i(f) = \frac{\sum_{x_j \in S_i} f(x_j)(k_i)}{\sqrt{\det(A_i)}}$$

where $S_i = \{x_j \in X \mid \Delta_{i,j}^\top A_i^{-1}\Delta_{i,j} < 1\}$ and A_i is the fuzzy shell covariance
matrix according to (7.1). (For circle clusters k_i with centres v_i and radii
r_i, we have $\Delta_{i,j} = (x_j - v_i) - r_i \frac{x_j - v_i}{\|x_j - v_i\|}$.)

Since small clusters have a low shell volume, they lead to high shell
densities. A *good* shell cluster k should therefore have not only a minimum
density but also a minimum number of data. Clusters that are evaluated
as *good* this way are temporarily ignored during the algorithm. The membership of a datum in a cluster is then computed from the distance to the
cluster contour. We delete the cluster data and store them in a temporary
data set. The data from the temporary data set will be considered again
later.

We regard two circles as compatible, if their centres as well as their radii are almost equal. An FCSS compatibility relation thus results from:

Definition 7.14 (Compatibility relation on FCSS clusters) *Let $p \in$ $\mathbb{N}$ and $C := \mathbb{R}^p \times \mathbb{R}$. For each $k_i = (v_i, r_i) \in C$, let v_i be the circle centre and r_i the radius of cluster k_i. For each $\gamma = (\gamma_1, \gamma_2) \in \mathbb{R}_+^2$ an FCSS compatibility relation $\dot{=}_\gamma \in C \times C$ is defined by*

$$\forall k_1, k_2 \in C: \quad k_1 \dot{=}_\gamma k_2 \quad \Leftrightarrow \quad ||v_1 - v_2|| \leq \gamma_1 \quad \wedge$$
$$|r_1 - r_2| \leq \gamma_2.$$

If we denote the smallest distance among the data as δ, we can use $\dot{=}_{(2\delta, 2\delta)}$. For the UFCSS algorithm, a number of heuristic parameters have to be chosen that are necessary to distinguish good and small clusters. Often, these values depend on the data, so different parameter sets have to be chosen for bit graphics, artificial test patterns, and two- or three-dimensional data, respectively. Consider a cluster $k_i \in K$ of a partition $f : X \to K$ and a datum $x_j \in X$.

- Cluster k_i obtains the predicate **"good"** if

$$\left(\sum_{j=1}^n f(x_j)(k_i) > \frac{|X|}{|K| + 1} \right) \quad \wedge \quad (SD_i > SD_{min}).$$

The minimum density for good clusters SD_{min} depends on the kind of data sets and should be chosen according to manually identified good clusters. The limit $\frac{|X|}{|K|+1}$ for the minimum data number was suggested by Krishnapuram in [67]. He assumes an approximately uniform distribution of the data over the clusters such that smaller clusters which were recognized correctly but have less than $\frac{|X|}{|K|+1}$ data points are not considered as *good* clusters.

- Datum x_j obtains the predicate **"belonging to cluster k_i"** if

$$||\Delta_{i,j}|| < \eta \sigma_i.$$

Here, η is a factor that is identical for all clusters, and σ_i is a measure for the thickness of the cluster or the scattering of the data around the cluster. In [67], $\eta = 3$ and $\sigma_i = \sqrt{\text{tr}(A_i)}$ were applied. (For the motivation of σ_i, cf. also page 216.)

• Cluster k_i obtains the predicate **"small"** if

$$\sum_{j=1}^{n} f(x_j)(k_i) < N_{tiny}.$$

Again, N_{tiny} depends on the kind and the number of data.

The following algorithm is slightly modified compared to the version in [67]. Only the general notions were used such that it can also be applied to other kinds of clusters and algorithms when the notions *small cluster*, *good cluster* etc. are specified.

Algorithm 5 (Unsupervised FCSS algorithm)

Given a data set X and a compatibility relation $\doteq$ on FCSS clusters.

Choose the maximum number of clusters c_{max}.
$c := c_{max}$, changes $:= false$, FirstLoop $:= true$
FOR 2 TIMES
    changes $:= false$
    REPEAT
        Run the FCSS algorithm with input (X, c).
        IF ($\exists$ clusters compatible with respect to $\doteq$) THEN
            Form transitive groups of compatible clusters.
            Merge each group to one cluster, update c.
            changes $:= true$
        ENDIF
        IF (FirstLoop=true) AND (there exist good clusters) THEN
            Deactivate good clusters and affiliated data,
            update c, changes $:= true$.
        ENDIF
        IF (there exist small clusters) THEN
            Delete small clusters, changes $:= true$, update c.
        ENDIF
    UNTIL ($c < 2$) OR (changes=false)
    FirstLoop $:= false$
    Reactivate all deactivated data and clusters, update c.
ENDFOR

In practical examples good results were obtained when c_{max} was at least 1.5 to 2.5 times the correct number of clusters. Moreover, the UFCSS algorithm needs a good initialization (e.g. by some FCM steps).

Let us consider the performance of the UFCSS algorithm first for the data sets from section 4.1 and 4.2. The results were very good without exception; all clusters were recognized correctly. (The results of the UFCSS algorithm for an unknown cluster number are better than those of FCSS with a known cluster number.) For the data set from figure 4.1, however, four circle clusters instead of three are recognized as optimal, but this only reflects the additional noise data, which were approximated by the fourth circle cluster. Behaviour like this was expected from the probabilistic version of UFCSS. Even the more complicated data sets from figures 4.5 and 4.6, respectively, are correctly recognized with $c_{max} = 20$. In this case an additional problem occurs for the very different data densities on the circle contours. An initialization with only 15 clusters led to just one FCM cluster for the three circles right on the top. On the other hand, the bold circles in the centre at the bottom and the two large circle segments were covered by many FCM clusters. These clusters, which are concentrated in a very narrow space, soon become similar FCSS clusters that are merged afterwards. In the area right on the top, clusters are thus missing for the correct determination of the cluster number.

However, if the cluster number is chosen too high, we can also run into problems: if c clusters approximate the same circle with n data, each datum obtains a membership $\frac{1}{c}$ in each of the c clusters. If the threshold value for a small cluster N_{tiny} is larger than $\frac{n}{c}$, all clusters are deleted. No cluster remains for this circle. This extreme case will hardly occur, however, if the compatible clusters are merged before small clusters are deleted (and we remember to recalculate the memberships).

Figure 7.23 shows a photograph of six washers from which a data set representing the edges of the depicted objects was extracted using image processing software. The result of the UFCSS algorithm for $c_{max} = 24$ is shown in figure 7.24. In two cases the algorithm detected intersecting clusters instead of concentric circles; a single cluster was found instead of two concentric circles twice; and one cluster is completely misplaced. The quality of the result strongly depends on the initialization, and thus also on the initial cluster number c_{max} as shown in figure 7.25, where we used $c_{max} = 20$. Although the FCM cluster providing the initialization for UFCSS cannot discover a whole circle correctly, every FCM cluster should at least detect a (single) circle segment. If the segments detected are long enough (e.g. consist of a certain amount of data vectors) the FCSS manages to approximate the corresponding circle in the first step. If such an initialization can be provided by the fuzzy c-means algorithm, the chances

Figure 7.23: Six washers

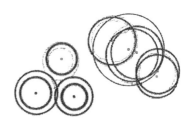

Figure 7.24: UFCSS analysis, $c_{max} = 24$, initialization using FCM

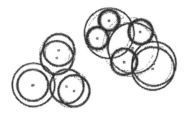

Figure 7.25: UFCSS analysis, $c_{max} = 20$, initialization using FCM

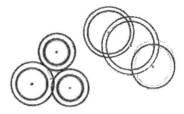

Figure 7.26: UFCSS analysis, initialization using edge detection

of a correct recognition of all circles are good. In the case of figure 7.23, however, the contours of two circles are quite close and adjacent. Each FCM cluster then covers not only one but at least two segments (of different circles). The subsequent FCSS run starts with the data of two circles as an initialization and is mostly not able to separate these data again. Supported by the non-Euclidean distance function, the minimization of the objective function leads in this case to clusters with rather small radii (relative to the original circles). If there are data inside the circle, the FCSS algorithm will use them as a stabilization of the cluster. Especially in the area in the middle on top of figure 7.25, the data are covered by several small and overlapping circles.

This can be avoided by a modified initialization. For the example in figure 7.23 we used an edge detection algorithm which does not merge the data of two adjacent circle contours and enables UFCSS to produce the result in figure 7.26. Even if the partition is not the optimum, it is clearly better than that obtained with an FCM initialization for $c_{max} \in \{20, 24, 28\}$. The edge detection algorithm is not sensitive to different cluster numbers as start values, since it estimates them by itself. (A description of the algorithm used for the edge detection can be found in section 7.3.)

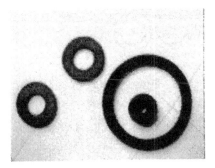

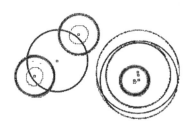

Figure 7.27: Four washers Figure 7.28: UFCSS analysis, $c_{max} = 15$

For good results it is important to distinguish between small and good clusters, leading to a simplification of the problem, i.e. a reduction of the number of clusters or data. Unfortunately, the validity measures and threshold values of the UFCSS algorithm are not very flexible. Figure 7.27 shows another picture from which a test data set for the UFCSS algorithm was generated. The sizes of the circles vary far more here than in the previous example. Even if the inner circles of the two washers on the left are correctly recognized, they do not reach the membership sum $\frac{|X|}{|K|+1}$ necessary for good clusters. Furthermore, we know from section 4.2 that circles enclosed by other circles are recognized better with an increased fuzzifier. However, the average membership decreases in this way, and the threshold value is then hardly reached even by large clusters.

Figure 7.28 shows the result of a UFCSS analysis that was initialized by 10 FCM steps ($m = 4$) and 10 FCSS steps ($m = 2$). Because of the FCM steps with a high fuzzifier, many FCM prototypes are moved to the circle centres and do not get close to the contours. These prototypes lead to a good approximation of the circle centres. In the course of the algorithm we first consider the nearest data points, thus $m = 2$ in the first FCSS steps. In the following steps, we need a more global view to separate circle contours, therefore $m = 4$ was chosen for the UFCSS algorithm. Compared to the fuzzifier the threshold value for the number of data had to be reduced to $\frac{|X|}{(|K|+1)(m-1)}$. The results cannot yet be considered as optimum because small and good clusters are not recognized as good clusters (e.g. the small inner circle) and the rather large circle in the middle of the picture is not identified as a bad one. Dense contours always result in rather good shell densities, even if a cluster only touches the actual contours in several places. Another problem is caused by the fact that the circles are often not ideal in digitized pictures but tend to a more elliptic shape (camera angle, shadows in the picture). For the large circle, this has been compensated by two circle clusters. The distance between the centres prevents the clusters from being merged here.

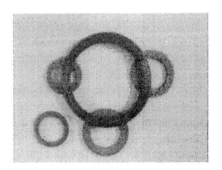

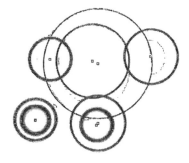

Figure 7.29: Four washers   Figure 7.30: UFCSS analysis

The technique for edge detection mentioned above, gives good results for the example in figure 7.28. The advanced initialization algorithm is necessary, especially when several large and small circle contours are overlapping, e.g. in figure 7.29. In this case, using an FCM initialization the FCSS algorithm covers all data vectors inside the large washer with many intersecting circles. Each cluster then *possesses* a contour segment of the large circle, which fixes its position. The local minimum, to which UFCSS converges, is destined by the initialization. An appropriate initialization by edge detection, on the other hand, leads to clearly better results from figure 7.30.

This result can be further improved by a more sophisticated cluster merging technique. For merging two clusters, the UFCSS algorithm checks whether the clusters represent the same circle by comparing their radii and centres. They are merged, only if the clusters are nearly identical. However, two clusters can also describe the same circle contour, even when they have clearly different centres and radii. This is the case when each cluster (poorly) approximates just one part of the same circle. In this case, further improvements can be obtained (for example in situations like on top of figure 7.25). Thus, alternatively merging methods from the U(M)FCQS can be borrowed (cf. section 7.2.4).

Finally, a third way to improve the partitions uses a simplification of the data sets as it is will be presented in section 7.3. However, this is no improvement of the UFCSS algorithm itself.

7.2.3   The contour density criterion

The local validity measures introduced so far are not satisfying from some points of view: dependent on the size and completeness of the contour, the values vary strongly, and thus further complicate the distinction between good and bad clusters. Since the predicate *good* can be applied to circles of different size as well as to full, half or quarter circles, reasonable threshold values which correctly separate all possible cases can often not be provided. The contour density criterion for shell clusters according to Krishnapuram, Frigui, and Nasraoui [66] compensates some of these weak points.

This local validity measure evaluates the data density along the cluster contour. It thus depends on the cluster size, since a high number of data is compensated by a longer extent of larger clusters. It has its maximum of 1 for an ideally closed (and may be partial) shell contour, i.e. a semi-circle also can have the maximum contour density 1. Hence, clear picture elements that have a densely marked contour but are partly hidden by other objects can be distinguished from poorly visible objects having only a sparsely marked contour. With the local validity measures so far, better values were often determined for sparse full circles than for dense quarter circles.

For the determination of the contour density of a cluster k_i we divide the number of data by the contour length. The number of data is computed as the sum of the memberships of all data vectors near enough to the shell contour:

$$S_i = \sum_{j \in Y_i} u_{i,j}, \quad Y_i = \{j \in \mathbb{N}_{\leq n} \mid \ \|\Delta_{i,j}\| < \Delta_{i,max}\}.$$

Here, $\Delta_{i,max}$ gives the maximum distance of a datum belonging to the contour. In the possibilistic case, we use $\Delta_{i,max} = \sqrt{\eta_i}$ because at this distance the memberships are 0.5 (for $m = 2$). The determination of the contour length is clearly more difficult. The contour length of a full circle with radius r is $2\pi r$, but the clusters often represent only a circle *segment*. If there is a closed and crisp contour, the length of a circle segment might be determined by tracking the data in the neighbourhood. But how long is the shell contour when the segment is only partially covered by data and the data only belong to a certain degree to the cluster? The validity measure of the contour density uses the covariance matrix (not the shell covariance matrix) of the cluster data with respect to the expected mean of the data (not the centre point of the geometrical figure). Let us first consider an ideal circle with a closed contour (i.e. an infinite number of data points). In this case, the mean value of the data vectors is the same as the centre of the circle. The covariance matrix A_i is $\begin{pmatrix} \frac{1}{2}r^2 & 0 \\ 0 & \frac{1}{2}r^2 \end{pmatrix}$. In the case of an ideal circle, we have $\mathrm{tr}(A_i) = r^2$.

Let us now consider the more general case of a circular arc with the same origin and radius. In order to simplify the determination of the covariance matrix, we assume that the circular arc with the angle γ is symmetrical to the x-axis, i.e. stretches from $-\frac{\gamma}{2}$ to $\frac{\gamma}{2}$. If we regard the data as realizations of a two-dimensional random variable Z, the covariance matrix is

$$
\begin{aligned}
A &= \mathrm{Cov}\{Z^2\} = E\{(Z - E\{Z\})^2\} \\
&= E\{Z^2\} - E\{Z\}^2 = \frac{1}{\gamma} \int_{-\frac{\gamma}{2}}^{\frac{\gamma}{2}} xx^\top \, d\varphi - mm^\top,
\end{aligned}
$$

where $x = (r\cos(\varphi), r\sin(\varphi))^\top$ is a point on the contour. The expected mean $m = E\{Z\}$ is

$$
m = \frac{1}{\gamma} \int_{-\frac{\gamma}{2}}^{\frac{\gamma}{2}} x \, d\varphi = \frac{1}{r\gamma} \begin{pmatrix} \int_{-\frac{\gamma}{2}}^{\frac{\gamma}{2}} r\cos(\varphi)d\varphi \\ \int_{-\frac{\gamma}{2}}^{\frac{\gamma}{2}} r\sin(\varphi)d\varphi \end{pmatrix} = \begin{pmatrix} \frac{2r}{\gamma}\sin(\frac{\gamma}{2}) \\ 0 \end{pmatrix}.
$$

Thus, we obtain

$$
A = \frac{r}{2} \begin{pmatrix} 1 - \frac{\sin(\gamma)}{\gamma} - \frac{8\sin^2(\frac{\gamma}{2})}{\gamma^2} & 0 \\ 0 & 1 + \frac{\sin(\gamma)}{\gamma} \end{pmatrix}.
$$

If we again compute the trace of the covariance matrix, we obtain

$$
\sqrt{\mathrm{tr}(A)} = r\sqrt{1 - \frac{4\sin^2(\frac{\gamma}{2})}{\gamma^2}}.
$$

In this case, the trace does not correspond exactly to the circle radius, as expected, but the difference is caused by the (square root) factor. Krishnapuram calls this radius, which is scaled with respect to the opening angle, the *effective* radius r_{eff} of the cluster and uses $2\pi r_{\text{eff}}$ as an approximation for the arc length. In the case of a full circle with $\gamma = 2\pi$, the equality $r_{\text{eff}} = r$ holds because $\sin(\pi) = 0$. The effective arc length is equal to the *real* length then. For a circular arc with angle γ and the contour length

Table 7.1: Compensation for partial contours

Value / Kind of cluster	Full circle	Semi- circle	Quarter circle	Straight line
$\frac{r_{\text{eff}}}{r}$	1.0	0.77	0.44	0.00
Contour density ϱ	1.0	0.65	0.57	0.55
Compensation factor f_P	1.0	1.54	1.74	1.81

L_γ, which can be interpreted as the number of data, we obtain the contour density

$$\varrho = \frac{L_\gamma}{2\pi r_{\text{eff}}} = \frac{\gamma}{2\pi\sqrt{1 - \frac{4\sin^2(\frac{\gamma}{2})}{\gamma^2}}}.$$

The density depends on the length of the circular arc (γ), but is independent of the cluster size (circle radius r). In order to compensate the influence of the arc length, we have to further modify the quality values. The expression $\frac{r_{\text{eff}}}{r}$ is suitable as a measure for the angle of the circle arc since it only depends on γ. From $\frac{r_{\text{eff}}}{r} = 1.0$, we can conclude that there is a full circle, from $\frac{r_{\text{eff}}}{r} = 0.77$ that there is a semi-circle and from $\frac{r_{\text{eff}}}{r} = 0.44$ that there is a quarter circle.

If the effective radius divided by the real radius suggests a half circle, we divide the density ϱ by the maximum density of a half circle (cf. table 7.1) and thus also obtain the maximum value 1 for partial contours. For the contour density criterion Krishnapuram distinguishes the cases full, half, quarter circle and straight line. He interpolates the scaling factor between these fixed items linearly. A straight line is considered as a special case of a circle with an infinite radius ($\frac{r_{\text{eff}}}{r} = 0$). For a straight line segment, the trace of the covariance matrix is identical to the variance $\frac{L^2}{12}$ where L is the length of the straight line segment. For the contour density $\varrho = \frac{L}{2\pi r_{\text{eff}}}$ we have $\frac{\sqrt{3}}{\pi} \approx 0.55$. The final local validity measure is defined in the following way:

Definition 7.15 (Contour density ϱ) *Let* $A(\mathbb{R}^2, R)$ *be an analysis space,* $X \subset \mathbb{R}^2$, $m \in \mathbb{R}_{>1}$ *a fuzzifier,* $f : X \to K \in A_{\text{fuzzy}}(\mathbb{R}^2, R)$ *a possibilistic cluster partition,* $u_{i,j} = f(x_j)(k_i)$ *the degree of membership and* $\Delta_{i,j}$ *the distance vector from datum* $x_j \in X = \{x_1, x_2, \ldots, x_n\}$ *and cluster* $k_i \in K = \{k_1, k_2, \ldots, k_c\}$. *As the (compensated) contour density* ϱ_i *of cluster* k_i *in the partition* f, *we use* $\varrho_i = \frac{S_i}{2\pi r_{\text{eff}}}$ *(or* $\varrho_i = \frac{S_i}{2\pi r_{\text{eff}}} \cdot f_P(\frac{r_{\text{eff}}}{r})$*),* *where*

$$S_i = \sum_{j \in Y_i} u_{i,j},$$

$$r_{\text{eff}} = \sqrt{tr(A_i)},$$

$$Y_i = \{j \in \mathbb{N}_{\leq n} \,|\, \|\Delta_{i,j}\| < \Delta_{i,\max}\},$$

$$A_i = \frac{\sum_{j=1}^{n} u_{i,j}^m (x_j - v_i)(x_j - v_i)^\top}{\sum_{j=1}^{n} u_{i,j}^m},$$

$$v_i = \frac{\sum_{j=1}^{n} u_{i,j}^m \, x_j}{\sum_{j=1}^{n} u_{i,j}^m} \quad and$$

$$f_P : \mathbb{R} \to \mathbb{R}, \, x \; \mapsto \; \begin{cases} 1.74 - 0.172(x - 0.44) & for \quad x \in [0.00, 0.44] \\ 1.54 - 0.594(x - 0.77) & for \quad x \in [0.44, 0.77] \\ 1.0 - 2.34(x - 1.0) & for \quad x \in [0.77, 1.00]. \end{cases}$$

Here, $\Delta_{i,\max}$ is the minimum distance from the contour of cluster k_i, where the data are not considered to belong to the cluster. It is recommended to choose $\Delta_{i,\max} = \sqrt{\eta_i}$.

If circle shaped contours are separated (for example by overlapping picture elements) into more than one circular arc, the contours do not obtain the maximum contour density even if the parts are ideal contours. A cluster, which approximates an arc, has high memberships in other arcs as well. The centre of gravity of the data is near the circle centre, and the effective radius is close to the actual radius. Hence, the correction factor for partial contours f_P will hardly cause any improvement of the contour density. When a contour is split into several parts there is no compensation by the validity measure, which might be interpreted as a weakness of the contour density. On the other hand, when the complete contour is subdivided into more and more segments, we deal with the transition to a sparsely marked contour – and in this case a decreasing measure is desirable.

In digitized pictures we have quantization effects that can distort the thickness, leading to wrong values of the validity measure. The number of data points S_i of a digitized ideal straight line k_i depends on its gradient. The validity measure assigns horizontal and vertical straight lines higher values than straight lines with other slopes. The straight line in figure 7.31 (gradient 1) has the length $L = \sqrt{2}$, the straight line in figure 7.32 (gradient $\sqrt{3}$) has the length $L = \frac{2}{\sqrt{3}}$ (if the pixel grid has the edge length 1). Despite the different lengths, both straight lines have 14 pixels. Horizontal and vertical straight lines in the pixel grid have the highest contour density, since they have the shortest length and the number of data points (14) remains constant. To be independent of the slope a correction factor f_D is introduced (similar to the one used above to compensate partial contours). For straight line segments with the angle φ this factor is

$$f_D = \frac{1}{\max\{|\cos(\varphi)|, |\sin(\varphi)|\}}. \tag{7.6}$$

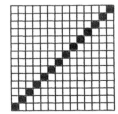

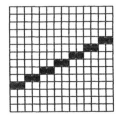

Figure 7.31: 45° straight line, $L \approx 1.41$

Figure 7.32: 30° straight line, $L \approx 1.15$

For all other kinds of clusters, we can imagine the contour as a compound of straight line segments. In this case, we correct each single datum on the contour depending on the vector tangent to the contour. If $\varphi_{i,j}$ is the angle between the horizontal axis and the vector tangent to the contour of cluster k_i at the position x_j, we replace the expression S_i from definition 7.15 by

$$\bar{S}_i = \sum_{j \in Y_i} \frac{u_{i,j}}{\max\{|\cos(\varphi_{i,j})|, |\sin(\varphi_{i,j})|\}}. \tag{7.7}$$

When the cluster is a straight line, the tangent vector and the horizontal axis intersect at the same angle as the straight line itself; the multiplication of the contour density with the factor f_D according to (7.6) is thus identical to the application of (7.7).

Krishnapuram, Frigui and Nasraoui [66] also consider theoretically the contour densities for ellipses, planes, spheres and ellipsoids. For some examples with ellipses they obtain satisfactory results using the non-compensated contour density corresponding to definition 7.15, when the main axis is much longer than the shorter axis. For the proportional value of 2, the contour density is very similar to that of circles. For a proportional value of 5, a correction via compensation mapping f_P is not necessary any more. (It would be unclear how both ellipse's radii could contribute to the compensation anyway.) We have to choose a proportional value between 2 and 5, from which on compensation is no longer necessary. To apply the proportion threshold in practice we have to know the ellipse radii, of course.

Remark 7.16 (Ellipse's radii) *If a two-dimensional, elliptic quadric is given by $p_1 x^2 + p_2 y^2 + p_4 x + p_5 y + p_6 = 0$, the ellipse's radii are*

$$r_1 = \sqrt{\frac{c}{p_1}} \quad and \quad r_2 = \sqrt{\frac{c}{p_2}},$$

where $c = \frac{p_4^2}{4p_1} + \frac{p_5^2}{4p_2} - p_6$.

Proof: We substitute $\bar{x} = x + \frac{p_4}{2p_1}$ and $\bar{y} = y + \frac{p_5}{2p_2}$ (quadratic completion) and have: $p_1 \bar{x}^2 + p_2 \bar{y}^2 = \frac{p_4^2}{4p_1} + \frac{p_5^2}{4p_2} - p_6$. We get the ellipse's radii from the equation in normal form: If r_1 and r_2 are the ellipse's radii, we have $\left(\frac{\bar{x}}{r_1}\right)^2 + \left(\frac{\bar{y}}{r_2}\right)^2 = 1$. The comparison of the coefficients leads to $r_1 = \sqrt{\frac{c}{p_1}}$ and $r_2 = \sqrt{\frac{c}{p_2}}$. ■

The simplified form of the quadric without the merged term $p_3 xy$, which is given in the remark, is obtained by rotating the ellipse as described in section 4.7. We can obtain values from the given formulae even if the quadric is *not* an ellipse. For the identification of the cluster types see A.3.

To determine the threshold values for a classification (e.g. in good clusters) we assume the maximum is one. In the practical application of the threshold values, it is necessary that the maximum can be reached. If several probabilistic clusters share the same contours, the memberships along the contours never favour a single cluster unambiguously, and thus the contour density will never have the value 1. The contour density criterion is therefore only suited for possibilistic cluster partitions. In addition, the contour thickness in a digitized picture should be exactly one pixel. With a *double* contour, the validity measure takes on values greater than 1. In this case the clusters may not be comparable and the threshold values may not be valid any more. A bad cluster, which crosses a thick contour often enough, can obtain a contour thickness close to 1. The results do not allow correct conclusions in this case. We cannot expect a useful analysis result. Instead, an algorithm for edge thinning [60] has to be applied.

Because each datum is weighted with its membership (maximum 1), each datum corresponds to a contour segment with the length 1. Even if we do not have a double contour, the contour densities might be falsified again, when the distance between the data vectors on the contour differs significantly from 1. If this is not the case, we have to use an additional factor incorporating the (average) smallest distance between two data vectors. This way, the data set is scaled to the 1×1 grid of a digitized picture, and therefore each datum corresponds to a contour segment of length 1.

7.2.4 The unsupervised (M)FCQS algorithm

In this section, the unsupervised FCSS algorithm is combined with the local validity measure of the contour density and the more general (M)FCQS algorithm. It is thus possible to recognize an unknown number of straight lines, circles and ellipses in a data set. Here, the basic principle of the UFCSS algorithm is not changed, but only the prototype-specific parts are adapted. Furthermore, the U(M)FCQS algorithm is different from the UFCSS algorithm because it uses possibilistic clustering.

Assume a cluster partition $f : X \to K$, a datum $x \in X$ and two clusters $k, k' \in K$.

- Cluster k obtains the predicate "small", if

$$\left(\sum_{x \in X} f(x)(k) < N_{VL} \right) \quad \vee \quad \left(\sum_{x \in X} f(x)(k) < N_L \wedge \varrho_k < \varrho_L \right).$$

A small cluster has either only very few data (membership sum smaller than $N_{VL} \approx \frac{2|X|}{100}$) or some more data but only a bad contour density (membership sum smaller than $N_L \approx \frac{4|X|}{100}$ and contour density smaller than $\varrho_L \approx 0.15$). In the second pass of the main loop of the U(M)FCQS algorithm, all clusters with membership sums smaller than N_L are regarded as small. With these stronger conditions for the first pass, we take into account that many clusters can better (i.e. with a lower shell thickness) approximate the data than few clusters.

- Cluster k obtains the predicate "good", if

$$(\varrho_k > \varrho_{VH}) \quad \vee \quad (\varrho_k > \varrho_H \wedge FHV_k < FHV_L).$$

Again, there are two possibilities for good clusters: either the contour density is very high ($\varrho_{VL} \approx 0.85$) or a somewhat lower density ($\varrho_L \approx 0.7$) is compensated by a very small fuzzy (shell) hypervolume ($FHV_L \approx 0.5$).

- Two clusters k and k' are "compatible", if

$$(T < T_L) \quad \wedge \quad (\varrho > \varrho_H).$$

Here, T and ϱ are the shell thickness and the contour density of the cluster resulting from merging the clusters k and k'. For this purpose all data with (possibilistic) memberships greater than α ($\alpha \approx 0.25$) in the clusters k or k' are determined and the possibilistic (M)FCQS algorithm is applied to these data. This complicated procedure is necessary since simple compatibility criteria cannot be provided for arbitrary quadrics as for the case of straight lines and circles. At least for the FCQS circle clusters we might want to use the FCSS compatibility relation. However, even for a data set consisting of circles and straight lines only, the algorithm may also detect other cluster shapes so that the complicated compatibility relation of the U(M)FCQS algorithm cannot be avoided.

As already mentioned, Krishnapuram, Frigui and Nasraoui transform hyperbolas, double straight lines and long-stretched ellipses directly into straight line clusters, before they start to merge all that clusters. The data belonging to the linear clusters are processed by the CCM algorithm with the straight line prototypes from the transformation. Straight line segments are thus merged with each other only, according to the CCM compatibility relation. They do not merge straight lines with other quadrics.

• The datum x obtains the predicate "belonging to cluster k", if

$$f(x)(k) > u_H.$$

For possibilistic algorithms, the memberships depend on the distance between the datum and the cluster only, so that a membership $u_{i,j}$ greater than $u_H = 0.5$ means the same as $||\Delta_{i,j}|| < \sqrt{\eta_i}$. For data sets from digitized pictures, however, values $\eta < 2$ make no sense, since the minimum distance is 1 (for horizontally and vertically adjacent pixels) and $\sqrt{2}$ (for diagonal pixels), respectively, because of the digitization. If the values of η drop below 2, all data vectors that do not belong unambiguously to the contour are classified as noise data. Hence, it makes sense to set $\eta_i = 2$ in general for the second possibilistic run. For the contour density determination the memberships always have to be computed with $\eta_i = 2$, too. (When applying the FCQS algorithm, the MFCQS distance function should be used for the contour density determination; the value $\eta_i = 2$ is chosen with respect to Euclidean distances. For the FCQS algorithm we should set $\eta_i > 2$, since the distance along the axes are different for an FCQS and MFCQS ellipse. Consequently, the accuracy of the computed ellipses varies (in terms of the Euclidean distance). This should be compensated by a higher value of η_i.)

• Datum x is called a "noise datum", if

$$\forall k \in K : \quad f(x)(k) < u_L.$$

If a datum is not correctly approximated, i.e. the memberships are all very small ($u_L \approx 0.1$), we assume a noise datum. Again, this threshold uses the fact of possibilistic memberships. With probabilistic clustering even a non-noisy data vector could obtain memberships smaller than u_L when this vector is approximated by many clusters.

In the previous definitions we provided standard values for the thresholds as used in [66]. For possibilistic algorithms it is easy to specify absolute values, because the memberships are not influenced by neighbouring clusters. In this way we can also classify noise data. The classification of noise data is very important since the relative number of noise data in the data set is increased by the (mostly incomplete) deactivation of good clusters and the corresponding data. If noise cannot be removed, the remaining clusters will not be recognized because of the more and more massive disturbances.

Compared to the UFCSS algorithm we have a different sequence for the elimination of small clusters. For the probabilistic memberships of UFCSS, the membership sum is small, even when several clusters share the same data. In order not to eliminate all clusters which approximate these data, small clusters are eliminated only after merging. For the possibilistic memberships of U(M)FCQS, this is no longer valid. If there really are several clusters sharing a contour, the membership sum of all clusters will be almost equal. The decision as to whether we deal with a small cluster can already be made in the beginning. By the elimination of small clusters, the number of clusters is decreased, which is good for the following merging operation. The merging operation of the U(M)FCQS algorithm is rather computationally complex, because the (M)FCQS algorithm has to be run for each pair of clusters. Since the number of the merged clusters that have to be tested depends on the cluster number quadratically, each missing cluster leads to a clear acceleration.

The deletion of small clusters at the end of the inner loop cannot be adopted to the U(M)FCQS algorithm from UFCSS, since a majority of data of other clusters might disappear with the elimination of good clusters. We cannot conclude then that there is a small cluster, because the orientation towards these data possibly happened only during the possibilistic part of the algorithm, perhaps because the originally approximated contour is too thin or or not sufficiently large. However, without the influence of data assigned to the good clusters the other clusters could have found a completely different orientation. From a low possibilistic membership sum after the elimination of good clusters, we can therefore not conclude that these clusters are superfluous.

For the three-dimensional case, Krishnapuram, Frigui and Nasraoui suggest a further modification of the (M)FCQS and U(M)FCQS algorithms.

Algorithm 6 (Unsupervised (M)FCQS algorithm)

Given a data set X and the maximum number c_{max} of clusters. $c :=$
c_{max}, changes := false, FirstLoop := true
FOR 2 TIMES
        changes := false
        REPEAT
                Run the algorithm with input (X, c).
                IF (there exist small clusters) THEN
                        Delete small clusters, changes := true, update c.
                ENDIF
                IF (there exist compatible clusters) THEN
                        Form groups of compatible clusters (cf. text).
                        Merge each group into one cluster,
                        changes := true, update c.
                ENDIF
                IF (FirstLoop=true) AND (there exist good clusters) THEN
                        Deactivate good clusters and affiliated data,
                        update c, changes := true.
                ENDIF
                IF (FirstLoop=true) AND (there exist noise data) THEN
                        Deactivate all noise data, changes := true.
                ENDIF
        UNTIL (c < 2) OR (changes=false)
        FirstLoop := false
        Reactivate all deactivated data and clusters, update c.
ENDFOR

On the one hand, the computationally expensive solution of high-degree
polynomials for the three-dimensional case is avoided by using the fuzzy
c-plano quadric shells (FCPQS) algorithm [66]. The algorithm only uses
approximated Euclidean distances, but memberships and prototypes are
computed on the basis of the same distance function (for MFCQS, the
Euclidean distance is only applied for the memberships, but not for the
computation of the prototypes). Furthermore, the infinite extension of the
clusters in the three-dimensional case gains even more influence. Therefore,
data of other clusters are often eliminated together with a good cluster, because the surface contours are intersecting. This causes gaps in the contours
of the remaining clusters, which makes the recognition more complicated.
In [66], further modifications can be found, which take these difficulties

into account. Here, we do not want to discuss the details neither for the three-dimensional case nor of the FCPQS algorithm.

In section 7.2.2 we used very strong conditions for the merging of two clusters. Only similar clusters were merged. So the number of clusters was reduced, but essentially the recognition performance was not influenced. With the U(M)FCQS algorithm, clusters can now be merged, which do not fulfil the strong conditions, as long as the union of the associated data results in a good cluster. This behaviour would have been desirable for the clusters in figure 7.25 (top middle view), too, because the large circle can possibly be recognized from several circles covering parts of the large circle contour. The basic idea of the (M)FCQS compatibility condition can also be adopted to the UFCSS algorithm, but here we do not choose the data for the merged cluster according to the memberships, but according to the distance from the circle contour. For U(M)FCQS the computationally expensive merging procedure is not that important, because the possibilistic memberships already help to avoid situations like those in figure 7.25. Another difference between the compatibility relations is that the (M)FCQS condition for compatible clusters only allows merging if a good cluster is produced; a unification of identical but bad clusters is not possible. Compared to the UFCSS algorithm, the essential compatibility criterion is not similarity but quality.

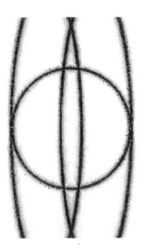

Figure 7.33: Three pairwise mergable clusters

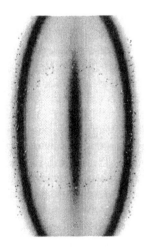

Figure 7.34: Resulting merged cluster

Different problems will occur during the application of the U(M)FCQS algorithm, depending on the kind of data sets. Sometimes we have to modify the algorithm to make the correct recognition possible. In the following we present some of these modifications. For example, we encountered another special case for merging clusters: if transitive groups of clusters are formed again, it may happen that pairs of clusters can be merged well, but do not produce a good new cluster altogether. This problem is caused by the application of the possibilistic algorithm. Let us consider a circle with two ellipse segments on its opposite sides stemming from the same ellipse and touching the circle. The circle and ellipse's segments are approximated by one cluster each (cf. figure 7.33). If we merge the circle cluster with an ellipse cluster, we may only obtain the circle cluster because of the majority of the circle data (longer contour) – and the circle cluster obtains a good contour density (possibilistic memberships). If we merge the two ellipse segments, the ellipse may be correctly recognized as the resulting merged cluster. All merged clusters have good contour densities and are thus pairwise mergable. However, if we merge all three clusters, the result does not contain a clear substructure with sufficient data. By merging all three clusters we lose all three correctly recognized clusters (cf. figure 7.34). In the worst case, the merging operation (with the MFCQS algorithm) determines a double straight line as the resulting cluster, which covers the ellipse segments by straight lines and the circle by (non-Euclidean) high memberships in the centre. In this case, we lose all the information about the clusters by merging. This example, which may seem to be somewhat artificial, shows that the compatibility relation does not give any information about the mergability of more than two clusters, except when similar new clusters are produced each time. Hence, we do not allow groups with more than two clusters and modify the original U(M)FCQS algorithm such that only a unification of pairs of compatible clusters is permitted. Since there is no obvious choice of the cluster pairs to be merged, we may prefer those with highest contour densities.

In practice, it can happen that merging two concentric circles yields only the larger circle: directly after merging, the cluster approximating the data of both circles is located in the middle between both contours; when applying possibilistic memberships, the contour moves to the larger circle (more data), and in the second possibilistic run, the cluster may be completely separated from the inner circle. (The FCQS algorithm is especially prone to this phenomenon, since its distance function gives lower values for data inside the contour, so that the objective function gives lower values when only the outer circle is considered as the resulting cluster.) We modify the UFCQS algorithm, so that formerly correctly recognized clusters do not vanish in the way described above: an additional condition for the successful merging of two clusters is a non-empty intersection of both contours before merging. As a measure for the intersection of two clusters k and k', we can use $k \cap k' := \sum_{x \in X} \min\{f(x)(k), f(x)(k')\}$: If there is a datum $x \in X$ which belongs to both considered clusters $f(x)(k)$ and $f(x)(k')$ will be close to 1 (possibilistic memberships). Then, the minimum is also close to 1, and datum x contributes to the sum. If x belongs to one contour only, the membership to the other contour will be close to 0, and the datum only influences the sum weakly. As a sufficiently large intersection, we can demand, for example, $k \cap k' > 5$. Here, the exact threshold value plays a rather inferior role; basically, only merging operations for completely disjunct contours should be avoided, where $k \cap k'$ is very small. Since this step can be computed faster than the merged cluster, the condition should be checked before the expensive merging operation. This way, the number of merged clusters that have to be computed is clearly reduced. If this simple condition is not sufficient, we can more generally require $k \cap k' > N_L \cdot \varrho$ where ϱ denotes the contour density of the merged cluster. This is a pessimistic approach, because we require a larger intersection for good result clusters. (For the FCQS algorithm, the possibilistic memberships with the actually determined η values should be used for the computation of the intersection. The restriction $\eta = 2$ holds for Euclidean distances only.)

As already mentioned in section 4.7, only the distances, and not the prototypes, are computed using the Euclidean distances with the MFCQS algorithm. We hope that MFCQS still minimizes the objective function in each step. This was not the case for some of the data sets tested, from time to time there were sudden deteriorations in the objective function, especially for larger data sets. Even if this caused only slight changes in the orientation of the prototypes, it prevented the algorithm from converging, since the memberships also reflect the sudden changes. This phenomenon may be caused by the limited computational precision when determining the Euclidean distances, i.e. the solution of the fourth-degree polynomials. However, it seems that the FCQS algorithm cannot always be *cheated* with memberships based on Euclidean distances, when the amount of data is increased, and may nevertheless converge. Sometimes, it takes a long time until the objective function reaches the level it had before the sudden change. Also, it may be that, after such a change, the algorithm converges to a local minimum of the objective function, which is higher than a formerly reached value. In our implementation, we therefore evaluated the objective function in each step and stored the present minimum. When a certain number of iteration steps did not yield an improvement of the present minimum, the iteration was interrupted and the present minimum was used as the analysis result. Although this modified interrupt condition makes the MFCQS algorithm applicable, it is not satisfying. The FCQS or the FCPQS algorithm should be used instead. Deteriorations in the objective function do not occur there, and the algorithms converge more rapidly. (In [66], the FCPQS algorithm is applied.) Applying the FCQS algorithm becomes somewhat complicated, because for the computation of the validity measure, we have to switch to the almost Euclidean MFCQS distances, but using the FCQS instead MFCQS is distinctly faster. (One hundred MFCQS iteration steps with 15 clusters and 1500 data require the solution of 2.25 million fourth-degree equations.) In the second possibilistic run, however, when using the FCQS the η values chosen should be a little greater than 2.

After the convergence of the FCQS iteration, the procedure for straight line recognition follows the original algorithm. Here, GK linear clusters are generated from long-stretched ellipses, double straight lines, hyperbolas and parabolas, and then the CCM algorithm is applied to these (cf. appendix A.4 for the recognition of FCQS straight line clusters). Only data with high memberships in the linear clusters are considered. In some cases, this procedure seems to be inappropriate. Often, complicated contours are covered by hyperbolas and double straight lines in just the start phase of the recognition. (This indicates a bad initialization.) This case also occurs when there are no straight lines in the data set. If we know that there are no straight lines in the data set, it does not make sense to run the FCQS line detection at all. However, the detection of a linear cluster can now be interpreted as a bad recognition performance. Figure 7.41 shows such a case. There are only ellipses and circles contained in the data set, however other cluster shapes are passing tangent to several circle contours. Large circles or ellipses can be well approximated at lower bendings by straight lines (or – even better – by hyperbolas or parabolas). If these clusters are evaluated as good clusters, the originally closed circle contour is segmented by the elimination of the data belonging to the linear clusters. This segmentation makes the recognition of large circles more complicated. (Moreover, in the original UFCQS algorithm, linear clusters and FCQS clusters are not merged, so that the segmentation is irreversible.)

However, if we know that no linear clusters are contained in the data set, we can convert them into point-shaped or circle-shaped clusters instead and let them find their new orientation. Without *resetting* these clusters, the linear contours would change only slightly and remain in a quite stable local minimum. If there are only circles contained in the picture, the UFCSS algorithm could be used instead of the CCM algorithm. However, we can also imagine a simple unsupervised FCM algorithm: the FCM clusters are merged if their distance is smaller than $\frac{N_L}{\pi}$. If we regard N_L as the smallest number of data for a cluster – since smaller clusters are eliminated – and imagine that these data form complete circles, we have $2\pi r \approx N_L$. The circle has the diameter $\frac{N_L}{\pi}$. Two FCM clusters sharing this circle can be assumed to be closer than this diameter. This defines a simple FCM compatibility relation. Which procedure gives the best results depends on the kind of the data sets. For strongly interlaced circles the FCSS algorithm provides only poor results by merging the data of different clusters. The FCM modification initializes the clusters for the next FCQS run in a more neutral way simply according to their neighbourhood.

Consider the example data set in figure 7.35. It demonstrates how the UFCQS algorithm works and which problems occur. Figure 7.36 shows the state of the UFCQS algorithm after some iterations of the inner loop. At that time, 17 clusters are still active. Figure 7.37 shows the data set after the elimination of small clusters, the merging (once) and extraction of good clusters (4 times). Notice the smaller ellipse cluster in figure 7.36 (bottom left), which was eliminated in figure 7.37 as a good cluster. The cluster approximated the inner circle in the lower left area, but it is reshaped to a flat ellipse by the other circle above. The right part of the circle data, which also belongs to the inner circle, is approximated by a different cluster. Although this cluster is not a complete circle, its (compensated) contour is high enough to be recognized as a good cluster. It would have been better if the cluster had remained in the data set and approximated the complete inner circle later, especially because the two other clusters have been merged there. Then, we could have hoped that the two other circles are separated into inner and outer circle again. The compensation of non-closed contours can thus also lead to a kind of *premature* classification as a good cluster.

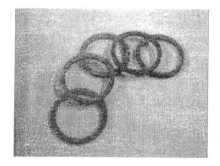

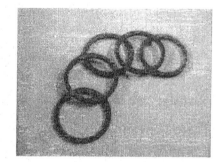

Figure 7.35: Five rings

Figure 7.36: The compensated contour density assigns a good density to non-closed contours, too

Figure 7.37: A cluster was too   Figure   7.38:   FCQS clusters
early evaluated as good   strongly respond to rest data

After the next FCQS iteration (figure 7.38), the merged cluster approximates the outer circle again. The remainders of the inner circle make sure
that a smaller cluster, from the ring located above it, is strongly drawn
downwards. This is an undesired effect, especially when the deformed cluster approximated a circle correctly before merging. Then the bad classification of the just deleted cluster also leads to the loss of another recognized
contour.

After that, the outer circle cluster is identified as a good cluster and
eliminated. There then remain only parts of the inner circle. FCQS clusters
are sensitive to these remainders. If several incomplete parts remain after
the extraction, they lead to many double straight lines, hyperbola and
parabola clusters. In figure 7.39, the two large and long-stretched ellipses
approximate on the one hand the remaining data, and on the other hand,
the two smaller contour pieces together with an already multiply covered
contour. The numerous intersections points with other contours, which
are also caused by the size of these clusters, prevent a classification as a
small cluster. However, the ellipses are not flat enough to be interpreted
as a linear cluster. They stay until they by chance (if at all) disappear
via a merging operation. The final analysis result is shown in figure 7.40.
The recognition performance is quite good, however some ellipse clusters
covering the contours twice could not be recognized as redundant.

The insufficient distance between data also had significant consequences
for the data set from figure 7.41 (for the original picture, cf. figure 7.27).
In the left part of the figure we have the remainders of a circle cluster
which was recognized as good and deleted. These remaining points are
approximated by numerous hyperbolas, parabolas and ellipses, although
they are noise data. Each of these clusters covers some other data segments

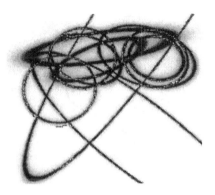 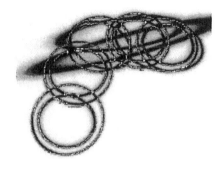

Figure 7.39: Sometimes, special *noise data clusters* are developed

Figure 7.40: UFCQS analysis

so that no cluster is *small*. Neither clusters nor noise data can be eliminated, since the clusters approximate the data well. The recognition can only be improved with a new orientation of the linear clusters. Although no straight line segments are contained in the figure, the special treatment of the linear clusters is thus justified. It takes care of the noise structure and gives the UFCQS algorithm a chance to recognize the correct clusters. Figure 7.42 shows the result from figure 7.41 when the linear clusters were eliminated and the unsupervised FCM mentioned above was applied to the corresponding data. Now, the few noise data are not covered at all any more because their number is not large enough for an FCM cluster of their own and the FCM clusters are not as *stretchable* as FCQS clusters. Before the next P-FCQS iteration, the noise data are recognized and eliminated. The former linear clusters are newly initialized and obtain a second chance for contour recognition.

 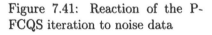 

Figure 7.41: Reaction of the P-FCQS iteration to noise data

Figure 7.42: Identification of noise data and resetting of the linear clusters

We can draw the following conclusion: noise and rest data massively disturb the recognition, and a new orientation with respect to badly covered data is necessary for pseudo-linear clusters. In order to avoid the generation of additional noise by incompletely removing cluster, the threshold values for the corresponding data can be decreased, so that more data near the clusters are deleted. Furthermore, a few MFCQS steps lead to a more precise adaptation to the data structure because of the Euclidean distance function. In addition, heuristic procedures for the elimination of unambiguous noise points can also be deployed. A new orientation of the linear clusters is necessary because of the tendency of the FCQS algorithm to cover whole areas of certain data sets by double straight lines, hyperbolas and parabolas. This is also the case for MFCQS clusters, which also create an extra region of high memberships between the straight lines of a double straight line cluster (cf. section 4.7). Thus, linear clusters are perfectly suited for *difficult* cases of minimizations with noise data as in figure 7.41. Since the (M)FCQS algorithm can not escape the local minimum – and furthermore, only approximates the noise data well – a new orientation of the linear clusters is necessary. For their new placement, data regions should be preferred that are badly covered by non-linear clusters. (These are not necessarily identical to those, which are approximated well by linear clusters.)

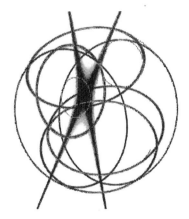
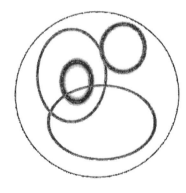

Figure 7.43: UFCQS analysis with FCM/FCSS initialization

Figure 7.44: UFCQS analysis with an initialization by edge detection

For more complicated data sets, the combination of FCM and FCSS can not always provide a suitable initialization. The data from figure 7.43 can not be recognized this way. The interlaced ellipses can only be covered by a network of circles. The UFCQS algorithm strictly sticks to initialization. However, if we use an algorithm for edge detection for the initialization, the FCQS algorithm finds the correct partition quickly (figure 7.44). For the UFCQS algorithm all available information about the data sets should be used for the initialization. Only in this way can results as in figures 7.45 and 7.46 be achieved.

Figure 7.45: Four washers

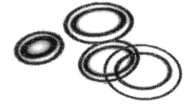

Figure 7.46: UFCQS analysis

7.3 Initialization by edge detection

For shell clustering we generally recommend an initialization different from that for compact clusters. Data can be very close to each other without belonging to the same contour. In cases like this, the algorithms FCM, GK and GG often merge data of different contours into one cluster and make the recognition difficult. Instead, we might use other clustering algorithms (e.g. FCSS); but this does not solve the initialization problem. In the literature, we find many clustering examples with explicitly mentioned *good initializations*, but it is not explained how these can be obtained. Perhaps, no comment is given on the initialization in order not to distract the reader from the brilliant results of the algorithm itself. For practical applications, however, a good initialization is at least as important as a good recognition algorithm. This applies especially, if the initialization can be achieved by relatively simple methods and decreases the run time of the UFCSS or U(M)FCQS algorithm. In this section we therefore introduce the technique for edge detection, which was used for some of the examples in this chapter.

We need information to specify a good initialization about the data to be partitioned. When we use the contour density as a validity measure, we implicitly assume very densely marked contours in our data (like in digitized pictures); otherwise, no cluster could obtain the predicate *good*. When data represent dense contours, the data set can easily be split into

Algorithm 7 (Edge detection)

Given a data set X, an environment radius r_{range} and $\gamma_1, \gamma_2 \in \mathbb{R}_+$.

FOR ALL $x \in X$
    Determine the environment of x:
        $U_x := \{y \in X \mid \|x - y\| < r_{\text{range}}\}$
    Determine the covariance matrix $C_x := \dfrac{\sum_{y \in U_x}(y-x)(y-x)^{\top}}{|U_x|}$

    Let n_x be the normalized eigenvector of C_x
    corresponding to the smallest eigenvalue
ENDFOR

FOR ALL $x \in X$
    Determine the homogeneity attribute:
    $h_x := \bigwedge_{y \in U_x} \left((|n_x^T n_y| > \gamma_1) \wedge (\,|(x-y)^T n_x| < \gamma_2) \right)$
ENDFOR

FOR ALL $x \in X$
    IF $h_x = true$ THEN
        Generate a new group set $G := \{x\}$
        $U := U_x,\ h_x := false$
        FOR ALL $y \in U \cap \{z \in X \mid h_z = true\}$
            $U := U \cup U_y,\ h_y := false,\ G := G \cup \{y\}$
        ENDFOR
    ENDIF
    IF $|G| \le 10$ THEN reject group G ENDIF
ENDFOR

---

connected disjoint components using a neighbourhood relation. A pixel x, for example, could be in the neighbourhood of a pixel y, if it is one of the 8 neighbour pixels of y (or if it is within another arbitrary maximum distance). When we form the transitive hull of this neighbourhood relation in our special data set, we obtain the disjoint connected components. These components of the data set from figure 7.28, for example, consist of circles; the data set from figure 7.44 is separated into the larger outer circle and the intersecting ellipses. In general, we obtain components, which have a decreased complexity with respect to the number of both data and clusters. The UFCSS and U(M)FCQS algorithms can then be applied to the components. Often, at least some of the simplified data sets are recognized at once. All components and recognized clusters are collected in the end, and the respective algorithm is applied again in the final step. Well-recognized components need not be included into that final run, if the recognized clusters show a very high contour density and a low shell thickness.

With the decomposition into connected components that hang together, the complexity of a data set is reduced but a correct edge detection is not necessarily achieved. Each component still has to be initialized before applying UFCSS or U(M)FCQS. However, if we decompose the data into such components, we can choose the following initialization: For each component, we run some fuzzy c-means steps. After that, the data vector (pixel) nearest to the FCM prototype is determined and (crisply) assigned to the cluster. Now, all neighbours of this pixel are also assigned to the corresponding cluster. The new assignment is repeated as long as all data are eventually distributed among the clusters. Since the data are connected, the algorithm terminates after a finite number of steps. The extracted contour is decomposed into parts of nearly the same size. The obtained partition can be used as an initialization for an analysis of the connected component, or it can be merged with the segments of all other components and serve as an initialization for the whole data set. Closely adjacent contours, which are not connected in the sense of the neighbourhood relation defined before, remain separate from each other in the new initialization. This would, for example, not be the case for a conventional FCM initialization. The recognition of data sets as in figure 7.28 is improved this way. If contours are intersecting, we still have data from different contours inside the same connected component.

For the examples in this chapter, we used further modifications, which are somewhat more flexible and general than the consideration of connected components. For each data vector (black pixel), we estimate the direction of the contour to which the pixel belongs. This is done by treating each data vector as a position vector of an FCV cluster. Using the data within a radius r_{range} around the position vector, we determine the orientation of the line so that the distance to the surrounding data is minimized. Therefore, we proceed as with the FCV algorithm, except that the position vector (cluster centre) is not computed, but each data vector acts as cluster centre, one after the other. Thus, for each data vector we obtain one direction vector. Let n_x be the vector perpendicular to the direction vector of datum x. Similar to the CCM algorithm, the small local linear clusters are compared, not to merge them but to build a chain. A datum $x \in X$ is said to be *homogeneously* in its r_{range}-environment, if for all data $y \in X$ with $\|x - y\| < r_{range}$ the condition $(|n_x^T n_y| > \gamma_1) \wedge (|(y - x)^T n_x| < \gamma_2)$ holds. This condition almost precisely corresponds to the first two conditions for compatible GK clusters according to the modified compatibility relation from definition 7.12. The missing third requirement considers the distance between the clusters, which can be dropped because the distances are limited to r_{range} anyway. In the sense of the modified CCM compatibility relation, a datum x which lies homogeneously in its environment is thus compatible with each local linear cluster in its environment. In contrast to the CCM algorithm, we do not merge clusters, but determine only the homogeneity attribute. Because the direction of the contour only changes slowly, all data of a circle or ellipse can lie homogeneously in their environment – without saying that all these local straight line segments have to be compatible with each other. If a datum x is near an intersection point of two contours, the data of the other contour get into the r_{range} environment, too, and the direction and normal vectors strongly deviate from the direction that was dominating before, as it can be seen in figure 7.47. Data close to intersection points are not homogeneous in their environment, connected areas of homogeneous data are separated by them. As the last step, these connected and homogeneous data are merged to groups. During the initialization of a shell clustering algorithm, each group is (crisply) assigned to a cluster. Each cluster obtains only data that are distributed almost uniformly along the contour of a single object (and possibly between two intersections with other objects). When the contour segments between the crossing points are long enough, the algorithm provides a very good initialization. For instance, the contours of the ellipses from figure 7.44 were separated by the edge detection algorithm only at the crossing points; each contour segment between the crossing points became a cluster. With this initialization, the UFCQS algorithm recognized the clusters after the first iteration step; then its only task was to eliminate the remaining clusters. As a nice side effect, the number of groups of connected and homogeneous data gives a good estimation for c_{max}, if very small groups are not counted.

For the choice of the threshold values we have to consider two contrary aspects. On the one hand, the threshold values have to be chosen as low as possible, in order not to accidentally connect the contours of two different

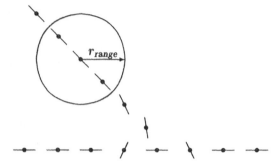

Figure 7.47: Local straight line pieces for edge detection

clusters at their intersection points. On the other hand, we do not want the contours to be too short, which can be the case when the thresholds are too low and the data are scattered around the ideal contour. If all data are gained by the same method, the contours will always be scattered to a similar extent, so that the parameters r_{range}, γ_1, and γ_2 only have to be adjusted once. The choice of the size of the data environment r_{range} depends on the distance between the contours of different clusters. If the contours are closer to each other than r_{range}, we will not obtain a large group of connected data. Here, partitioning the data vectors into the above-mentioned connected components makes sense, because the contours can be separated from each other then.

In the examples of this chapter, the parameters $(r_{range}, \gamma_1, \gamma_2) = (5\delta, 0.9, 2\delta)$ were used, where δ is the (average) smallest distance between data pairs. For digitized pictures, δ was not computed (e.g. according to 5.15) but set close to 1. For the choice of the parameter r_{range} we should also consider the size of gaps in the contours. If the gaps in the digitized contours are longer than r_{range}, they cannot be assigned to the same edge during edge detection. A value of $r_{range} = 5$ already closes small gaps in the contour.

As already mentioned, a strict choice of parameters leads to short contour segments. For the FCSS algorithm, these short circular arcs are sufficient to recognize the actual circle, while the FCQS algorithm approximates groups like these by linear clusters. For the UFCQS algorithm, only weak homogeneity conditions should be chosen, or it should be initialized by a few FCSS steps. It is also useful to improve the digitized pictures using standard image processing techniques. For example, isolated noise points can be deleted and contours can be thinned for the application of the contour density criterion [60].

Chapter 8

Rule Generation with Clustering

This chapter shows how the membership matrices obtained from fuzzy clustering algorithms can be extended to continuous membership functions. These membership functions can be used to describe fuzzy if-then rules. We distinguish the generation of fuzzy rules for classification systems and for function approximators.

8.1 From membership matrices to membership functions

The cluster algorithms described in the previous chapters determine prototype locations, prototype parameters, and a matrix of memberships. This matrix contains the memberships of the elements of the data set in each of the clusters. If the data set is representative for the system it comes from, we can assume that additional data cause only slight modifications of these clusters. If we can neglect these modifications, we might want to determine the memberships of these additional data without running through the whole cluster algorithm again. In general, we might want to determine the memberhips of all possible data (e.g. $\mathbb{R}^p$). To achieve this we have to extend the discrete membership matrix to a continuous membership function. In this section we describe some approaches for this extension.

239

8.1.1 Interpolation

Consider the data set

$$X = \left\{ \left(\frac{1}{7}, \frac{6}{7} \right), \left(\frac{2}{7}, \frac{3}{7} \right), \left(\frac{3}{7}, \frac{5}{7} \right), \left(\frac{4}{7}, \frac{2}{7} \right), \left(\frac{5}{7}, \frac{4}{7} \right), \left(\frac{6}{7}, \frac{1}{7} \right) \right\} \subset [0,1]^2$$

(8.1)

shown in figure 8.1. After random initialization of the cluster centres we used the fuzzy c-means algorithm (Euclidean norm, $m = 2$, $c = 2$, 100 steps) and obtained the cluster centres

$$V \approx \{(0.69598, 0.30402), (0.30402, 0.69598)\} \tag{8.2}$$

and the membership matrix

$$U \approx \begin{pmatrix} 0.07826 & 0.28099 & 0.06198 & 0.93802 & 0.71901 & 0.92174 \\ 0.92174 & 0.71901 & 0.93802 & 0.06198 & 0.28099 & 0.07826 \end{pmatrix}.$$

(8.3)

If we assign each data point to the cluster in which it has the largest membership, we obtain the cluster partition shown in figure 8.1: Cluster 1 contains the points x_4, x_5, x_6, and the points x_1, x_2, x_3 belong to cluster 2.

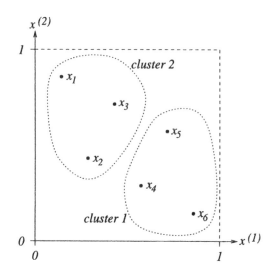

Figure 8.1: The data set X

Figure 8.2 shows the cluster centres as solid circles and the memberships of the data to the second cluster as "telephone poles" parallel to the u axis.

We consider the tips of the discrete membership poles u_{ik}, $i = 1, 2$, $k = 1, \ldots, 6$, as samples of continuous membership functions $\mu_i : \mathbb{R}^2 \to [0, 1]$, $i = 1, 2$.

$$\mu_i(x_k) = u_{ik}, \quad i = 1, 2, \quad k = 1, \ldots, 6. \tag{8.4}$$

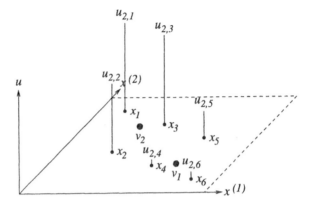

Figure 8.2: Cluster centres v_i and memberships $u_{2,k}$ in the second cluster for the data set X

One way to obtain these continuous functions is to connect the discrete memberships using linear interpolation, as shown in figure 8.3. The continuous membership functions define the membership values of the test vectors $x = (x^{(1)}, x^{(2)}) \in [0, 1]^2 \backslash X$ which are not in the training set. Notice that the interpolation only yields memberships of points inside the convex hull of the data set.

8.1.2 Projection and cylindrical extension

Membership functions can often be assigned linguistic labels. This makes fuzzy systems transparent, i.e. easy to read and interpret by humans. In one-dimensional domains the labels "low", "medium", or "high" are frequently used. It is often difficult to specify meaningful labels for membership functions with higher dimensional domains. For example, how would you call the membership function in figure 8.3? Assigning labels is often easier in one-dimensional domains. We therefore project the memberships u_{ik} from (8.3) to the $x^{(1)}$ and $x^{(2)}$ axis, respectively. Figure 8.4 shows these projections for the membership values in the second cluster. We apply linear interpolation (as described in the previous section) to the projected

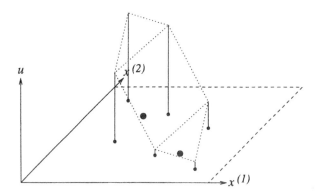

Figure 8.3: Linear interpolation of memberships

memberships and obtain the dotted membership functions. In our example, the membership function over the domain $x^{(1)}$ can be assigned the label "low" or "$x^{(1)}$ is low", and the membership function over $x^{(2)}$ can be called "high" or "$x^{(2)}$ is high".

The elements of our data set X in (8.1) are two-dimensional. The membership functions obtained by linear interpolation of the projected memberships are defined for one-dimensional arguments only. We can simply extend these functions to multi-dimensional arguments by considering only one of the argument components. This so-called cylindrical extension means that the projection of the multi-dimensional argument vector is used as the scalar argument of the membership function in the projection space. The membership function is extended to all other dimensions like a "profile". This profile specifies the membership of a two-dimensional vector $x = (x^{(1)}, x^{(2)})$ in the fuzzy sets "$x^{(1)}$ is low" and "$x^{(2)}$ is high", respectively.

With these labels we can interpret the original cluster as "$x^{(1)}$ is low and $x^{(2)}$ is high", i.e. the conjunction (="and") of the cylindrical extensions ("$x^{(1)}$ is low", "$x^{(2)}$ is high"). In fuzzy set theory, the conjunction is realized using a t-norm, e.g. the minimum. The dotted three-dimensional object in figure 8.4 shows the minimum of the two profiles "$x^{(1)}$ is low" and "$x^{(2)}$ is high." It is a continuous membership function which satisfies the interpolation condition (8.4). It is a linguistically interpretable alternative to the direct (high dimensional) linear interpolation shown in figure 8.3.

The crisp set of elements with the membership greater than α is called α-cut. If we cut through the linear interpolated projection to the $x^{(1)}$ axis in figure 8.4, we obtain non-convex α-cuts, e.g. for $\alpha \approx 0.8$. We therefore call this membership function *non-convex as a fuzzy set*, i.e. it contains non-convex α-cuts. Also the linear interpolated projection to the $x^{(2)}$ axis and the conjunction of the cylindrical extensions in figure 8.4 are non-convex as fuzzy sets.

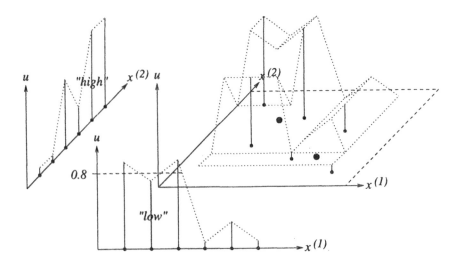

Figure 8.4: Projected and linear interpolated memberships and the conjunction of their cylindrical extensions

8.1.3 Convex completion

In the previous section we presented a method to obtain continuous membership functions by (a) projection of the memberships to the coordinate axes, (b) linear interpolation of the projected memberships, (c) cylindrical extension of the membership functions in the projection space, and (d) conjunction of the cylindrical extensions. The advantage of this method in comparison with the linear interpolation in the product space (figure 8.3) is a higher transparency. The membership functions can be assigned linguistic labels and are therefore easier to interpret by humans. They are, however, non-convex as fuzzy sets as shown in figure 8.4. The resulting irregular surface might not be desirable in many applications. Therefore we now try to approximate the non-convex fuzzy sets with convex membership functions. This approximation is called *convex completion* [97]:

We denote $\xi_1, \dots, \xi_k$, $\xi_1 \leq \dots \leq \xi_k$, as the ordered projections of the vectors $x_1, \dots, x_k$ and $\nu_{i1}, \dots, \nu_{ik}$ as the respective membership values. To obtain the convex completion we eliminate each point (ξ_t, ν_{it}), $t = 1, \dots, k$, for which two limit indices $t_l, t_r = 1, \dots, k$, $t_l < t < t_r$, exist, so that $\nu_{it} < \min\{\nu_{i\,t_l}, \nu_{i\,t_r}\}$. After this convex completion we obtain continuous convex membership functions by linear interpolation of the remaining points (ξ_t, ν_{it}). Figure 8.5 shows the result of convex completion for our example. If convex completion is involved, the interpolation condition (8.4) will generally not hold any more.

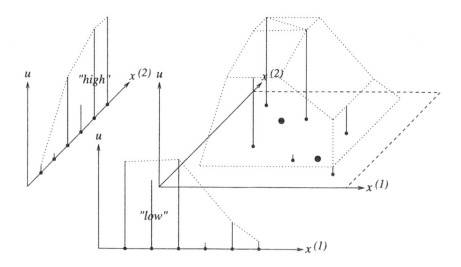

Figure 8.5: The conjunction of the cylindrical extensions after convex combination

The methods for the generalization of membership matrices described so far are summarized in table 8.1. Concerning the processing steps the method from section 8.1.2 is an extension of section 8.1.1 and the method from section 8.1.3 extends both of them. Concerning the properties, however, none of them can be considered superior. The choice depends on the application and the demanded properties.

8.1.4 Approximation

An alternative approach to extend the discrete memberships to continuous membership functions uses parametric membership function prototypes.

Table 8.1: Methods to extract membership functions and their properties

section	8.1.1	8.1.2	8.1.3
figure	8.3	8.4	8.5
processing steps:			
projection		×	×
convex completion			×
linear interpolation	×	×	×
cylindrical extension		×	×
conjunction		×	×
properties:			
interpolation condition	×	×	
linguistically interpretable		×	×
convexity	×		×

By adapting the parameters of these prototypes, the membership function is fitted to the membership values. This adaptation can be done by minimizing the (e.g. quadratic) approximation error.

A straightforward method is the use of the membership function prototype used in the clustering algorithm itself. The fuzzy c-means alternating optimization, e.g. computes membership values as

$$u_{ik} = 1 / \sum_{j=1}^{c} \left(\frac{\|x_k - v_i\|_A}{\|x_k - v_j\|_A} \right)^{\frac{2}{m-1}}, \quad i = 1, \dots, c, \quad k = 1, \dots, n. \quad (8.5)$$

The discrete membership values obtained by FCM can therefore be interpolated using the respective continuous FCM membership function

$$\mu_i(x) = 1 / \sum_{j=1}^{c} \left(\frac{\|x - v_i\|_A}{\|x - v_j\|_A} \right)^{\frac{2}{m-1}}, \quad i = 1, \dots, c. \quad (8.6)$$

Figure 8.6 shows the approximation of the memberships in the second cluster from our example (8.3). Notice that in this special case the approximation is an interpolation. Since the u_{ik} are calculated using (8.5), they fit *exactly* to the surface (8.6). Figure 8.6 also shows that the FCM membership function is not convex as a fuzzy set.

A similar membership function can be obtained from the possibilistic c-means alternating optimization with the membership update equation

$$u_{ik} = \frac{1}{1 + \left(\frac{\|x_k - v_i\|_A}{\sqrt{\eta_i}} \right)^{\frac{2}{m-1}}}, \quad i = 1, \dots, c, \quad k = 1, \dots, n. \quad (8.7)$$

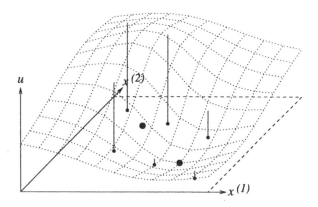

Figure 8.6: Interpolation with FCM membership function

This can be generalized to the PCM membership function

$$\mu_i(x) = \cfrac{1}{1 + \left(\cfrac{\|x - v_i\|_A}{\sqrt{\eta_i}}\right)^{\frac{2}{m-1}}}, \quad i = 1, \ldots, c. \tag{8.8}$$

The t_1 or Cauchy distribution is defined as

$$F_{\text{Cauchy}}(t) = \frac{1}{2} + \frac{1}{\pi} \arctan t, \tag{8.9}$$

$t \in \mathbb{R}$. The t_1 or Cauchy density function is

$$f_{\text{Cauchy}}(t) = \frac{1}{\pi} \frac{1}{1 + t^2}. \tag{8.10}$$

For $m = 2$, $\eta_i = 1$, Euclidean distance $\|x\| = \sqrt{x^T x}$ and $p = 1$, the PCM membership function (8.8) is

$$\mu_i(x) = \pi \cdot f_{\text{Cauchy}}(x - v_i). \tag{8.11}$$

We therefore call the PCM membership function a *Cauchy function*. A generalized Cauchy function is defined as

$$\mu_i(x) = \cfrac{1}{1 + \left(\cfrac{\|x - v_i\|}{\sigma_i}\right)^{\alpha}} \tag{8.12}$$

with the shape parameter $\alpha \in \mathbb{R}$ and the spread parameters $\sigma_i \in \mathbb{R}^+\backslash\{0\}$. The (generalized) Cauchy function, and in particular the PCM membership function are convex as a fuzzy set. Other common convex prototypes are the Gaussian membership function

$$\mu_i(x) = e^{-\left(\frac{\|x - v_i\|}{\sigma_i}\right)^{\alpha}} \tag{8.13}$$

and the extended hyperconic membership function

$$\mu_i(x) = \begin{cases} 1 - \left(\frac{\|x - v_i\|}{\sigma_i}\right)^{\alpha} & \text{for } \|x - v_i\| \leq \sigma_i \\ 0 & \text{otherwise.} \end{cases} \tag{8.14}$$

Figures 8.7 and 8.8 show the approximation of the second cluster from (8.3) with a Gaussian function and a cone, respectively. Notice that the Gaussian membership function is always nonzero and the cone is restricted to a closed circle of radius σ_2 centred at v_2.

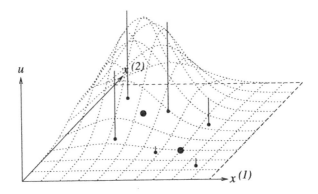

Figure 8.7: Approximation with Gaussian function, $\sigma_2 = 1/4$, $\alpha = 2$

In the previous examples we approximated the discrete membership values with a continuous membership function in the *product space*. According to the methods in the sections 8.1.2 and 8.1.3, we can also approximate the memberships in the *projection space*. Figure 8.9 shows the approximation of the projected memberships from our example with triangles and the pyramid representing the conjunction of the cylindrical extensions. In the literature, also trapezoidals are often used as prototypes for this method instead of triangles [51].

8.1.5 Cluster estimation with ACE

In the previous sections we used an objective function-based clustering algorithm like FCM-AO to determine a membership matrix describing the

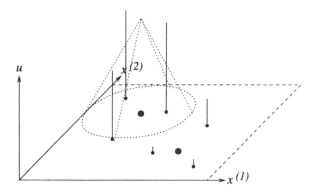

Figure 8.8: Approximation with cone, $\sigma_2 = 1/4$, $\alpha = 1$

cluster partition. The discrete memberships were then extended to a membership function by interpolation or approximation with membership function prototypes. If the desired membership function prototype is known in advance, it can already be directly implemented in the clustering algorithm. In chapter 6 this method was called *alternating cluster estimation (ACE)*. When the clusters are determined using ACE, the determination of continuous membership functions is very simple. We can just take the membership function prototypes with the selected parameters at the positions given by the cluster centres in V. In contrast to the approximation method described in the previous section, no curve fitting or additional optimization is necessary.

The projection methods used with the other approaches to membership function extraction can also be applied to cluster estimation with ACE. In this case, we can use the *projections* of the membership function prototypes with the selected parameters at the positions given by the *projected* cluster centres in V.

8.2   Rules for fuzzy classifiers

The continuous membership functions obtained with the methods from the previous sections describe fuzzy clusters. In this chapter we show how these clusters can be associated with fuzzy classes, and how rules for fuzzy classifier systems can be built.

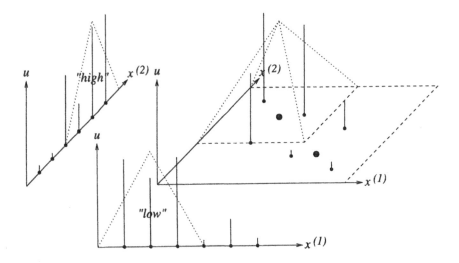

Figure 8.9: Approximation with conjunction of triangles, $\sigma_2 = 1/4$, $\alpha = 1$

8.2.1 Input space clustering

In the previous section we discussed the aspect of lingustic interpretability and assigned labels (like "low" and "high") to clusters. This means that the elements of the clusters belong to *classes* which are characterized by the corresponding labels. In this sense clusters can be interpreted as classes. The clustering algorithms described in this book can therefore be used to build classifiers. Note that classical fuzzy clustering algorithms compute fuzzy sets that are closely related to the fuzzy sets used in fuzzy rule bases [68]. These fuzzy sets can be interpreted in terms of a reprentative value – similar to a cluster centre – and a scaling of the distance [50, 51]. The membership degree of datum to the corresponding fuzzy set decreases with increasing scaled distance to the reprentative value. This interpretation of fuzzy rules applies to classification, as well as to function approximation and control [52].

Let us first consider classifiers built by *unsupervised learning*: the labels are not contained in the data set, but can be assigned after clustering. Even if the labels are accessible in advance, it is often easier to apply unsupervised techniques, if the (usually large) data set has to be labelled manually. Consider the example of a traffic jam detection system: an induction loop in the road is used as a metal detector which serves as a sensor for cars moving across or standing on it. Each passing car causes a pulse from the sensor. The frequency f of the pulses is equal to the number of cars per minute and the pulse width is inversely proportional to v, the current speed of the car. After some prepocessing procedures we obtain a data set containing the features car frequency f and speed v. For each feature vector we could manually provide the additional information, if there is a traffic jam or not, and apply supervised techniques. Unsupervised clustering, however, is much easier in this case. After unsupervised clustering we consider the cluster centres, and decide which of them is a typical traffic jam and which is not. Using this unsupervised approach only a few manual classifications are necessary.

We have to distinguish between the *training data set* which is used for clustering and building the classifier, and the *test data set* which is afterwards classified without influencing the clusters. Given a feature vector from the test data set, the classifier determines its memberships in all the classes. These memberships can be interpreted as degrees of compatibilities with the class prototypes. The output values of the classifier are $\mu_i(f, v)$, $i = 1, \ldots, c$, the memberships in each cluster. Each output can be interpreted as the result from a *classifier rule*. If in our example the prototype of the first cluster is interpreted as a traffic jam situation, we can write the cluster as a corresponding rule

$$\text{if } \mu_1(f, v) \text{ then ``traffic jam''.}$$

The classifier output $\mu_1(f, v)$ is a direct result of this rule. Given a feature vector this classifier rule is used to determine a membership quantifying to which extent we have a traffic jam. Some applications use only the label of the class with the highest membership as the classifier output. We call this a *defuzzified* result.

8.2.2   Cluster projection

We have mentioned previously that it is often difficult to assign linguistic labels to membership functions with high dimensional domains. Projections offer a higher transparency and interpretability. Therefore we used membership functions in the projection space and reconstructed the high dimensional membership function of the clusters as the conjunction of the convex completions. We can also apply this technique for the generation of transparent classifier rules. If we denote the membership functions from our example in the projection space as $\mu_i^{(f)}(f)$ and $\mu_i^{(v)}(v)$, $i = 1, \ldots, c$, respectively, we can write the traffic jam classifier rule as

$$\text{if } \mu_1^{(f)}(f) \text{ and } \mu_1^{(v)}(v) \text{ then ``traffic jam''.}$$

If we assign labels to the membership functions we finally obtain the readable rule

R_1 : "if car speed is low and car frequency is low then traffic jam".

It should be stressed that this rule in general does not yield the same results as the original rule with multi-dimensional membership functions. It is an approximation like the projection techniques described in the previous sections.

Potential of fuzzy classifiers

For many classification problems a final unique assignment of an object under conideration to a class is required. For fuzzy classification systems as they are considered above, the unique assignment corresponds to a defuzzification process which simply chooses the class with the highest membership degree. One might ask the question what the advantages of fuzzy rules over crisp rules for classification problems are.

First of all, in addition to the crisp classification the additional information about the membership degrees to the classes is available. Looking at the differences between these membership degrees provides some hints as to whether the final decision for a single class for the considered object was unique or ambiguous.

Another advantage is the greater flexibility of this classification calculus in comparison to a calculus based on crisp sets. In order to see in which sense fuzzy classification rules are more flexible let us consider crisp classification rules in the sense that the fuzzy sets $\mu_R^{(i)}$ are assumed to be characteristic functions of crisp sets, say intervals. ($\mu_R^{(i)}$ denotes the fuzzy set for the ith variable appearing in rule R.) Then it is obvious that in the two-dimensional case, each rule can be associated with the rectangle induced by the two intervals appearing as characteristic functions in the if-part of the rule. Therefore, a classifier based on crisp sets, carries out the classification by partitioning the data space in rectangular regions.

A classification problem with two classes that are separated by a hyperplane, i.e. a line in the two-dimensional case, is called linearly separable. Obviously, a linearly separable classification problem can be solved only approximately by crisp classification rules by approximating the separating line by a step function (see figure 8.10).

For fuzzy classification rules the situation is much better. In the two-dimensional case, classification problems with two classes that are separated by a piecewise monotonous function can be solved exactly using fuzzy classification rules [55]. Before we can prove this result, we have to introduce the notation that we need to formalize the problem.

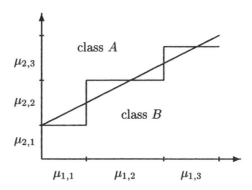

Figure 8.10: Approximate solution of a linearly separable classification problem by crisp classification rules

We consider fuzzy classification problems of the following form. There are p real variables $x_1, \ldots, x_p$ with underlying domains $X_i = [a_i, b_i]$, $a_i < b_i$. There is a finite set $\mathcal{C}$ of classes and a partial mapping

$$\text{class} : X_1 \times \ldots \times X_p \longrightarrow \mathcal{C}$$

that assigns classes to some, but not necessarily to all vectors $(x_1, \ldots, x_p) \in X_1 \times \ldots \times X_p$.

The aim is to find a fuzzy classifier that solves the classification problem. The fuzzy classifier is based on a finite set $\mathcal{R}$ of rules of the form $R \in \mathcal{R}$:

R: If x_1 is $\mu_R^{(1)}$ and $\ldots$ and x_p is $\mu_R^{(p)}$ then class is C_R.

$C_R \in \mathcal{C}$ is one of the classes. The $\mu_R^{(i)}$ are assumed to be fuzzy sets on X_i, i.e. $\mu_R^{(i)} : X_i \longrightarrow [0, 1]$. In order to keep the notation simple, we incorporate the fuzzy sets $\mu_R^{(i)}$ directly in the rules. In real systems one would replace them by suitable linguistic values like *positive big*, *approximately zero*, etc. and associate the linguistic value with the corresponding fuzzy set.

We restrict ourselves to max-min rules here, i.e. we evaluate the conjunction in the rules by the minimum and aggregate the results of the rules by the maximum. Therefore, we define

$$\mu_R(x_1, \ldots, x_p) = \min_{i \in \{1, \ldots, p\}} \left\{ \mu_R^{(i)}(x_i) \right\} \qquad (8.15)$$

as the degree to which the premise of rule R is satisfied.

$$\mu_C^{(\mathcal{R})}(x_1,\ldots,x_p) = \max\{\mu_R(x_1,\ldots,x_p) \mid C_R = C\} \qquad (8.16)$$

is the degree to which the vector $(x_1,\ldots,x_p)$ is assigned to class $C \in \mathcal{C}$. The defuzzification – the final assignment of a unique class to a given vector $(x_1,\ldots,x_p)$ – is carried out by the mapping

$$\mathcal{R}(x_1,\ldots,x_p) = \begin{cases} C & \text{if } \mu_C^{(\mathcal{R})}(x_1,\ldots,x_p) > \mu_D^{(\mathcal{R})}(x_1,\ldots,x_p) \\ & \text{for all } D \in \mathcal{C}, D \neq C \\ unknown \notin \mathcal{C} & \text{otherwise.} \end{cases}$$

This means that we finally assign the class C to the vector $(x_1,\ldots,x_p)$ if the fuzzy rules assign the highest degree to class C for vector $(x_1,\ldots,x_p)$. If there are two or more classes that are assigned the maximal degree by the rules, then we refrain from a classification and indicate it by the symbol *unknown*. Note that we use the same letter $\mathcal{R}$ for the rule base and the induced classification mapping.

Finally,

$$\mathcal{R}^{-1}(C) = \{(x_1,\ldots,x_p) \mid \mathcal{R}(x_1,\ldots,x_p) = C\}$$

denotes the set of vectors that are assigned to class C by the rules (after defuzzification).

Lemma 8.1 *Let* $f : [a_1,b_1] \longrightarrow [a_2,b_2]$ $(a_i < b_i)$ *be a monotonous function. Then there is a finite set* $\mathcal{R}$ *of classification rules to classes* P *and* N *such that*

$$\begin{aligned} \mathcal{R}^{-1}(P) &= \{(x,y) \in [a_1,b_1] \times [a_2,b_2] \mid f(x) > y\}, \\ \mathcal{R}^{-1}(N) &= \{(x,y) \in [a_1,b_1] \times [a_2,b_2] \mid f(x) < y\}. \end{aligned}$$

Proof. Let us abbreviate $X = [a_1,b_1]$, $Y = [a_2,b_2]$. Define the fuzzy sets

$$\mu_1 : X \longrightarrow [0,1], \qquad x \mapsto \frac{b_2-f(x)}{b_2-a_2},$$

$$\mu_2 : X \longrightarrow [0,1], \qquad x \mapsto \frac{f(x)-a_2}{b_2-a_2} = 1 - \mu_1(x),$$

$$\nu_1 : Y \longrightarrow [0,1], \qquad y \mapsto \frac{y-a_2}{b_2-a_2},$$

$$\nu_2 : Y \longrightarrow [0,1], \qquad y \mapsto \frac{b_2-y}{b_2-a_2} = 1 - \nu_1(y).$$

The fuzzy sets are illustrated in figure 8.11. The rule base consists of the two rules

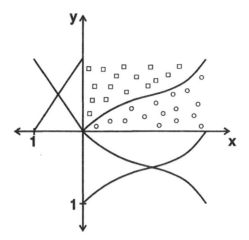

Figure 8.11: The fuzzy sets for the classification rules

R_1:   If x is μ_1 and y is ν_1 then class is N.

R_2:   If x is μ_2 and y is ν_2 then class is P.

- Case 1. $f(x) > y$.

  – Case 1.1. $\mu_1(x) \geq \nu_1(y)$.
  This implies $\mu_2(x) \leq \nu_2(y)$. Thus we have

$$\mu_N^{(\mathcal{R})}(x,y) \;=\; \nu_1(y)$$

$$=\; \frac{y - a_2}{b_2 - a_2}$$

$$<\; \frac{f(y) - a_2}{b_2 - a_2}$$

$$=\; \mu_2(x)$$

$$=\; \mu_P^{(\mathcal{R})}(x,y).$$

  – Case 1.2. $\mu_1(x) < \nu_1(y)$.
  Thus we have $\mu_2(x) > \nu_2(y)$ and

$$\mu_N^{(\mathcal{R})}(x,y) \;=\; \mu_1(y)$$

$$= \frac{b_2 - f(x)}{b_2 - a_2}$$

$$< \frac{b_2 - y}{b_2 - a_2}$$

$$= \nu_2(x)$$

$$= \mu_P^{(\mathcal{R})}(x, y).$$

- Case 2. $f(x) < y$ can be treated in the same way as case 1 by replacing $<$ by $>$.

- Case 3. $f(x) = y$.
  This means $\mu_1(x) = \nu_2(y)$ and $\mu_2(x) = \nu_1(y)$, i.e. $\mu_N^{(\mathcal{R})}(x, y) = \mu_P^{(\mathcal{R})}(x, y)$. ∎

It is obvious that we can extend the result of this lemma to piecewise monotonous functions, simply by defining corresponding fuzzy sets on the intervals where the class separating function is monotonous (see figure 8.12) and defining corresponding rules for each of these intervals so that we have the following theorem.

Theorem 8.2 Let $f : [a_1, b_1] \longrightarrow [a_2, b_2]$ $(a_i < b_i)$ be a piecewise monotonous function. Then there is a finite set $\mathcal{R}$ of classification rules to classes P and N such that

$$\mathcal{R}^{-1}(P) = \{(x, y) \in [a_1, b_1] \times [a_2, b_2] \mid f(x) > y\},$$
$$\mathcal{R}^{-1}(N) = \{(x, y) \in [a_1, b_1] \times [a_2, b_2] \mid f(x) < y\}.$$

A direct consequence of lemma 8.1 and its proof is that we can solve two-dimensional linear separable classification problems with only two fuzzy classification rules incorporating simple triangular membership functions so that we are in a much better situation than in the case of crisp classification rules.

Unfortunately, theorem 8.2 does not hold for higher dimensions – at least when we restrict ourselves to max-min rules [55]. Nevertheless, the two-dimensional case shows that a fuzzy classifier can solve a classification problem using a very simple and compact representation.

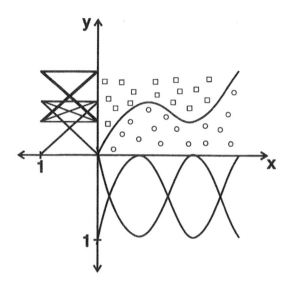

Figure 8.12: The fuzzy sets for the classification rules of a piecewise monotonous function

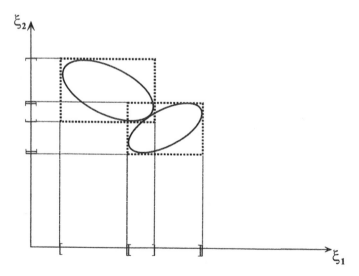

Figure 8.13: Information loss from projection

Information loss from projection

The classification rule derived from a fuzzy cluster represents an approximation of the cluster, but not the exact cluster. The value

$$\mu_{R_i}(\xi_1, \ldots, \xi_p) = \min_{\alpha \in \{1, \ldots, p\}} \left\{ \mu_i^{(\alpha)}(\xi_\alpha) \right\}$$

to which the premise of the R_i is satisfied for a given datum $(\xi_1, \ldots, \xi_p)$ and the membership degree of the datum to the corresponding cluster are not equal in general. One reason for this is that the projections of the clusters are usually only approximated by corresponding fuzzy sets. But even if we had the exact projections of a cluster, we lose information when we only consider the projections of a (fuzzy) set. The cluster shape of FCM clusters is spherical in the ideal case. But projecting an FCM cluster in order to obtain a classification rule means that we consider the smallest (hyper-)box that contains the corresponding (hyper-)sphere. In the case of the Gustafson-Kessel and the Gath-Geva algorithms the clusters are ellipses or ellipsoids in the ideal case. The projections of an ellipse do not allow us to reconstruct the ellipse, but only the smallest rectangle that contains the ellipse (figure 8.13). The loss of information can be kept small, when we restrict to axes-parallel ellipses or ellipsoids. When we are interested in constructing rules from fuzzy clusters, the simplified versions of the GK and GG algorithms in section 2.4 are a good compromise between flexibility and minimization of the loss of information. They are more flexible than FCM, but avoid the larger loss of information of GK and GG, since the loss of information is relatively small for axes-parallel ellipses.

Axes-parallel clusters

Figure 8.14 shows a set of rules that were generated from fuzzy clusters of the simplified version of the Gath-Geva algorithm. We used a a training data set with 150 data with four attributes. Each datum belongs to one of three classes. We allowed up to 5% misclassifications. The rules classify three data incorrectly and three more are not classified at all. This effect is caused by the approximation of the projection of the fuzzy clusters by trapezoidal fuzzy sets. Trapezoidal fuzzy sets have a bounded support so that we cannot avoid the fact that the projections are cut off at some point. Data beyond these cut off values can not be classified by the corresponding rule, although they might have a non-zero membership in the cluster. We can avoid this effect by using Gaussian or other types of convex membership functions instead of trapezoidal functions for the approximation of the projections.

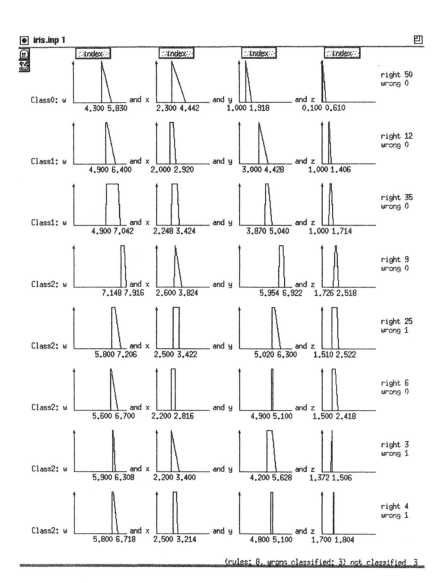

Figure 8.14: Classification rules induced by fuzzy clustering

Since we have to accept a certain loss of information, when we project fuzzy clusters, we could interpret the rules just as visualizations of the fuzzy clusters, but nevertheless use the memberships in the clusters for the classification. Or we can also optimize the fuzzy sets by other methods – for instance using neuro-fuzzy techniques as they are described in [79]. These methods lead to further improvements [78].

Grid clusters

In [53] grid clustering was proposed in order to completely avoid the loss of information. Grid clustering is restricted to well-behaved triangular membership functions (in the sense that the membership degrees at each point add up to 1), and aims at finding fuzzy partitions for the single domains on the basis of multi-dimensional data. For the grid clustering we assume that we are given a data set $\{x_1, \ldots, x_n\} \subset \mathbb{R}^p$. We are looking for fuzzy partitions on the single domains consisting of triangular membership functions with the restrictions that for each domain, at most two supports of different fuzzy sets have a non-empty intersection, and the sum of the membership degrees is one at any point of the domain. This means that we have to determine a suitable grid in the multi-dimensional space. We assume that for each domain the number of triangular membership functions is predefined. We define the membership degree of a data point to a cluster represented by a grid point as the minimum of the membership degrees of the triangular membership functions whose tips are the projections of the grid point. The grid clustering algorithm introduced in [53] is not based on an objective function, but relies on a heuristic strategy for constructing the clusters.

In order to improve this grid clustering algorithm, a suitable objective function that can be optimized by an evolutionary algorithm was proposed in [54].

The aim of this strategy is to place the prototypes on a suitable grid in the multi-dimensional space in order to get fuzzy partitions on the single domains consisting of triangular membership functions. For this reason our objective function should not depend on the order in which we analyse the single dimensions. The coordinates of the prototypes should only be influenced by those coordinates of the data that are relatively near (in this case near means a small Euclidean distance) to the prototype's coordinate. A simple way in single dimensions is to let the data between two prototypes only influence these two. On a grid there are a lot of prototypes with the same coordinates in a single dimension. In the objective function, each coordinate has to be taken into account only once. These considerations led us to the following objective function:

$$J = \sum_{r=1}^{p}\sum_{j=1}^{k_r}\left(\sum_{\substack{s\in\{1,\dots,n\}: \\ c_{j-1}^{(r)}<x_{sr}<c_{j}^{(r)}}} \left(\frac{c_j^{(r)} - x_{sr}}{c_j^{(r)} - c_{j-1}^{(r)}} \right)^m \right.$$

$$\left. + \sum_{\substack{\ell\in\{1,\dots,n\}: \\ c_{j-1}^{(r)}<x_{\ell r}<c_{j}^{(r)}}} \left(\frac{x_{\ell r} - c_j^{(r)}}{c_{j+1}^{(r)} - c_j^{(r)}} \right)^m \right).$$

x_{sr} and $x_{\ell r}$ are the r-th coordinate of the data x_s and x_ℓ, respectively ($s,\ell \in \{1,\dots,n\}$). We assume that we have in the r-th dimension ($r \in \{1,\dots,p\}$) k_r triangular fuzzy sets. Each triple $c_{j-1}^{(r)}, c_j^{(r)}, c_{j+1}^{(r)}$ induces a triangular membership function with the interval $(c_{j-1}^{(r)}, c_{j+1}^{(r)})$ as the support and $c_j^{(r)}$ as the point with the membership degree one. Thus the fractions in the sums (without the power m) provide the value one minus the membership degree to the triangular membership function with the tip at $c_j^{(r)}$ of the data (or better: their r-th projection) that lie on the support of the membership function. Since we add up the values for all triangular fuzzy sets (and the sum of the membership degrees of a datum to neighbouring fuzzy sets yields one), we obtain the smallest considerable value of J, when all the memberships degree are 0.5 (as long as $m > 1$ is chosen) and the largest considerable value, when the membership degrees are either zero or one. We want the $c_j^{(r)}$ to be in the centre of data clusters, i.e. the membership degree is high (near one) for data in the cluster and low (near zero) for data in other clusters. Thus we aim at maximizing J. Note that J is a measure very similar to the partition coefficient.

A special treatment is needed for the data on the left/right of the leftmost/rightmost prototype (the values $c_1^{(r)}$ and $c_{k_r}^{(r)}$). In the beginning, we assume that the $c_j^{(r)}$ are uniformly distributed, i.e. equidistant. We add in each dimension two additional prototypes $c_0^{(r)}$ and $c_{k_r+1}^{(r)}$, again equidistant to the left and right of $c_1^{(r)}$ and $c_{k_r}^{(r)}$. The values $c_0^{(r)}$ and $c_{k_r+1}^{(r)}$ are assumed to be fixed and must not be changed during the optimization process. Nevertheless, we have to take these additional prototypes into account in the objective function so that the data at the edge of the universe have the same influence as the data in the middle. This means that the second sum in J actually goes from $j = 0$ to $j = k_r + 1$. For the construction of the prototypes we only need the grid coordinates in each dimension.

Since the objective function of the grid clustering algorithm is not differentiable, we cannot apply an alternating optimization scheme. Therefore, the objective function of the grid clustering algorithm is directly optimized by an evolutionary algorithm [4, 75]. We cannot discuss the details of evolutionary algorithms here. An overview on the application of evolutionary algorithms to fuzzy clustering in general is provided in [54].

8.2.3 Input output product space clustering

In sections 8.2.1 and 8.2.2 we have restricted ourselves to unsupervised clustering, where the class labels are not available or not used in the training step. This unsupervised training scheme can also be used for supervised classification: Using the class labels the data set is divided into one set for each class, and these classes are then clustered separately. In this way multiple clusters are obtained for each class. All the clusters of each class can be translated into one disjunctive rule like

$$\text{if } \mu_2(f, v) \text{ or } \mu_3(f, v) \text{ then "light traffic"},$$

where $\mu_2(f, v)$ and $\mu_3(f, v)$ are the membership functions generated for the two clusters in the data set for the class label "light traffic". The membership functions can be projected to the domain components and assigned linguistic labels as presented in section 8.2.2. We obtain the readable rule

$$R_2 : \text{"if car speed is medium and car frequency is high}$$
$$\text{or car speed is high and car frequency is low}$$
$$\text{then light traffic"}.$$

The resulting classification system can be completely specified by the rules R_1, R_2, and the corresponding membership functions $\mu_1^{(f)}(f)$, $\mu_2^{(f)}(f)$, $\mu_3^{(f)}(f)$, $\mu_1^{(v)}(v)$, $\mu_2^{(v)}(v)$, and $\mu_3^{(v)}(v)$.

8.3 Rules for function approximation

In the last section we presented methods to build fuzzy *classification* systems from fuzzy clusters. In this section similar methods are derived for the generation of fuzzy *function approximation* systems.

8.3.1 Input ouput product space clustering

When dealing with classifiers we interpreted the data set X as a set of feature vectors. Now we want to consider the problem to approximate a function f from given samples X. In this case X becomes a set of pairs of input vectors x_k and output vectors $f(x_k)$. The data set X (8.1) that we used throughout this whole section can be considered for example as noisy samples of the function

$$f(x) = \frac{21}{4} - \frac{x}{2}, \quad x \in [0, 1], \tag{8.17}$$

whose graph is plotted as the dashed line in figure 8.15. If we apply the

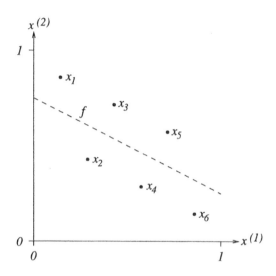

Figure 8.15: The function f described by the noisy samples X

same methods as in sections 8.1.1 (interpolation), 8.1.2 (projection and cylindrical extension), or 8.1.3 (convex completion), we obtain the clusters which were already shown in figures 8.3, 8.4, or 8.5, respectively. Although all of these methods can be used for function approximation we focus here – as a representative example – on the cluster generation method with convex completion described in section 8.1.3. The classifier rule obtained there was written as

"if $x^{(1)}$ is low and $x^{(2)}$ is high then class 2".

The classes can be interpreted as parts of the graph of the function f. The corresponding function approximation rule for the second local enviroment on the graph (or class 2) can therefore be equivalently written as

R_3 : "if $x^{(1)}$ is low then $x^{(2)}$ is high".

In this interpretation the class labels become "typical values" in local graph environments.

Mamdani-Assilian systems

How do we evaluate a function approximation rule like R_3? Mamdani and Assilian [71] suggested the algorithm shown in figure 8.16.

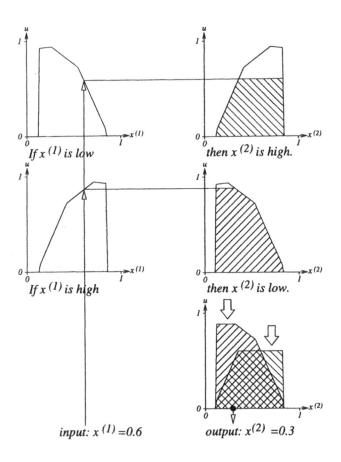

If $x^{(1)}$ is low   then $x^{(2)}$ is high.

If $x^{(1)}$ is high   then $x^{(2)}$ is low.

input: $x^{(1)} = 0.6$ output: $x^{(2)} = 0.3$

Figure 8.16: Mamdani-Assilian rule evaluation

For each rule the firing or activation grade is calculated as the value of the membership function of the "if" part (left-hand side, LHS) for the given input $x^{(1)}$. The fuzzy result of each rule is the area under membership functions of the "then" part (right-hand side, RHS) which is cut at the height given by the firing grade (shaded areas). The fuzzy results are aggregated to a fuzzy result of the whole rule base by overlapping all the RHS areas (white arrows). The fuzzy output is converted into a crisp output $x^{(2)}$ by a so-called defuzzification operation [84], which is often implemented as the centroid of the shaded area. For multi-dimensional inputs the membership functions can be projected to the single domains and connected with "and". In this case the firing value of each rule is calculated as the t-norm (e.g. minimum) of the memberships of each of its LHS components.

Singleton systems

The evaluation of Mamdani-Assilian systems can be simplified by de-fuzzifying the RHS membership functions before the aggregation of the rule results. These defuzzified RHSs are crisp values, which can be represented in the fuzzy set context as vertical poles, so-called singletons. A singleton rule R_i with the crisp RHS value y_i is written as

$$\text{if } \mu_i(x) \text{ then } x^{(p)} = y_i,$$

where $\mu_i(x)$ is the firing grade of rule R_i given the input x. The firing grade can be a single membership value or a (e.g. conjunctive) combination of several LHS membership values.

If the rule results are aggregated by addition and defuzzified using the centroid method, the output $x^{(p)}$ of a singleton system is

$$x^{(p)} = \frac{\sum_i \mu_i(x) \cdot y_i}{\sum_i \mu_i(x)}. \tag{8.18}$$

If the singletons are the centroids of the RHS membership functions of a Mamdani-Assilian system with the same LHS structure, both systems yield the same input output behaviour. For maximum aggregation and centroid defuzzification the singleton system has a different transfer function, if the RHSs overlap.

To obtain RHS singletons from the partition matrix, we have to perform projection, convex completion, linear interpolation, and defuzzification. A much easier approach simply takes the projections of the cluster centres to the output space as the RHS singletons. This simple method leads to different, though in our example more reasonable, results, as we will see later.

Takagi-Sugeno systems

Singleton systems are not only special cases of Mamdani-Assilian systems, but also special cases of Takagi-Sugeno systems [98]. Takagi and Sugeno replaced the crisp values y_i in singleton systems with functions f_i of the input x and obtained rules like

$$\text{if } \mu_i(x) \text{ then } x^{(p)} = f_i(x),$$

which are – for sum aggregation and centroid defuzzification – evaluated according to (8.18) as

$$x^{(p)} = \frac{\sum_i \mu_i(x) \cdot f_i(x)}{\sum_i \mu_i(x)}. \tag{8.19}$$

These Takagi-Sugeno systems approximate a function f in the local environment $\mu_i(x)$ with the local approximator $f_i(x)$.

The RHS functions $f_i(x)$ of Takagi-Sugeno systems can be obtained from clustering with higher order prototypes. Linear functions for first order Takagi-Sugeno systems can be obtained from the FCV [103], FCE [93, 85], or Gustafson-Kessel [39, 76, 2, 49] algorithm. The position and angle of each prototype are used to define the offset and the system matrix of the local linear functions $f_i(x) = A_i \cdot x + b_i$.

The input output characteristics of the previously described four system architectures for the approximation of the function f (8.17) are shown in figure 8.17:

1. a Mamdani-Assilian system with minimum inference, maximum aggregation, and centroid defuzzification,

2. a singleton system, where the RHSs are the defuzzified RHS membership functions from the Mamdani-Assilian system (equivalent to a Mamdani-Assilian system with minimum inference, additive aggregation, and centroid defuzzification),

3. a singleton system, where the RHSs are the projections of the cluster centres to the output space, and

4. a first order Takagi-Sugeno system obtained by fuzzy c-elliptotypes clustering with $\alpha = 0.9$.

In this example the best approximation was obtained with the singleton system with projected cluster centres (3). Due to the relatively high noise level the first order Takagi-Sugeno system (4) produced the worst approximation.

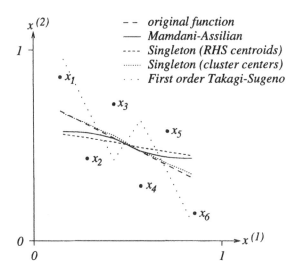

Figure 8.17: Function approximation with different fuzzy system models

8.3.2   Input space clustering

Fuzzy function approximators can also be generated with clustering in the input space only. The projection of X to the input space is then used as the data set, and membership functions are generated using the methods described in section 8.1. These membership functions represent the LHSs of Mamdani-Assilian systems, singleton systems, or Takagi-Sugeno systems. To determine the corresponding RHSs the whole data set has to be processed again.

Mamdani-Assilian systems

The front diagram in figure 8.18 shows the vertical poles representing the memberships obtained from input space clustering. The arrows indicate how these memberships are associated with the points of the original data set in the input output product space. The poles in the input output space can be projected to the output space (horizontal arrows) and postprocessed using interpolation, convex completion, approximation, and label assignment. The resulting membership function (left diagram in figure 8.18) is appropriate as RHS for a Mamdani-Assilian system. The LHS and RHS in figures 8.18 and 8.5 are not much different. For our example clustering in the input space yields almost the same result as clustering in the input output product space.

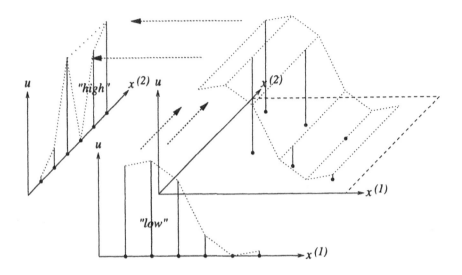

Figure 8.18: Generation of RHS memberships by projection

Singleton systems

Using the methods described above we can use results from input space clustering to generate RHS membership functions for Mamdani-Sugeno systems. According to the methods for input output clustering we can also defuzzify these RHS membership functions (e.g. with the centre of gravity method) and obtain RHSs for singleton systems. In contrast to input output clustering we do not have cluster centres in the input output product space here whose projections could be used directly as RHS singletons.

Takagi-Sugeno systems

To generate first order Takagi-Sugeno systems we have to determine the offsets and regression matrices for the linear RHS functions. To calculate the offsets b_i we consider the approximated functions f_i at the cluster centres v_i. We assume that the $f_i(v_i)$ are the defuzzified RHSs d_i which also served above as RHSs for singleton systems and obtain

$$f_i(v_i) = A_i \cdot v_i + b_i = d_i \quad \Rightarrow \quad b_i = d_i - A_i \cdot v_i. \qquad (8.20)$$

To determine the local system matrices A_i we suggest two methods:

1. Define crisp clusters by assigning each data point to the cluster in which it has its largest membership. Then determine the system matrix by linear regression.

2. For each cluster compute the within cluster fuzzy scatter matrix as described in chapter 3, and calculate the linear RHS coefficients from the matrix eigenvectors.

Higher order Takagi-Sugeno systems can be generated using nonlinear regression methods.

8.3.3   Output space clustering

Output space clustering yields no reasonable results for classification problems. For function appoximation, however, it can be applied according to input space clustering: The memberships in the output space are associated with the points in the input output product space and projected to the input space leading to LHS membership functions. The memberships in the output space themselves can be used to generate RHS membership functions for Mamdani-Assilian systems. These RHS membership functions can be defuzzified to obtain RHSs for singleton systems. Also the cluster centres in the output space can be used as singleton RHSs. This is in contrast to input space clustering. The singleton systems can be extended to first order Takagi-Sugeno systems by computing the offsets and local system matrices with one of the two methods described above for input space clustering. Nonlinear regression can be used accordingly to generate higher order Takagi-Sugeno systems.

8.4   Choice of the clustering domain

For the generation of Mamdani-Assilian function approximation systems we obtained nearly the same results for input output clustering (figure 8.5). and for input clustering (figure 8.18). In general the rule extraction methods always produce similar classifiers and function approximators for all three domains, input space, output space, and input output product space.

For classification problems input output clustering is possible only for labelled data. If labels are available, it seems advantageous to use all information, and therefore use the whole input output product space. For labelled *classification* data clustering in the output space leads to no reasonable results. All three domains can be used for *function approximation*.

Singleton systems with cluster centre projections as the RHS (which lead to good approximations) cannot be generated by input space clustering. Which domain is the best for clustering depends on the structure of the data, the properties of the application, and on the target system architecture. It is still an open question [81]. At least one of the various methods described in this chapter, however, should lead to satisfactory solutions for most applications.

Appendix

A.1   Notation

Certain symbols, which have the same meaning in the whole text, are listed in this appendix. Also, the notations used are explained in table A.2, except for well-known standard notations.

All figures of cluster partitions contain in their caption the name of the algorithm that produced the results. This does not apply to figures which show the same data set in different versions or states. Here, the applied algorithm is mentioned in the text, while the caption refers to the differences with respect to the partition in this case. The algorithms are written with the usual abbreviations, which are also listed in the index. The figures in chapters 2 and 3 always illustrate probabilistic memberships, unless there is a $P-$ in front of the algorithm's name. For the sake of clarity, probabilistic partitions are presented by possiblistic memberships in chapter 4 and the following chapters, even if there is no $P-$ in the caption. There, $P-$ indicates that the results were obtained by possibilistic clustering. For the possibilistic presentation of probabilistic clustering results, a constant value was chosen for all η values. For the differences between a probabilistic and a possibilistic partition, cf. chapter 1.

A.2   Influence of scaling on the cluster partition

Not all of the analysis results, which can be found with the algorithms introduced above, are independent from the scaling of the data set to be analysed. However, this is the case for the memberships. If a scaled distance function is used, the scaling factor cancels out for the memberships corresponding to theorem 1.11 because of the computation of relative distances. In the possibilistic case, the extension coefficients η adapt to the

271

Table A.2: Symbols and denotations

$(a_i)_{i \in \mathbb{N}_{\leq l}}$	abbreviation for an l-tuple $(a_1, a_2, \ldots, a_l)$
$\equiv$	$f \equiv x$ denotes the constant function that maps all values to x
$\|x\|_2, \|x\|$	Euclidean norm of x
$\|x\|_p$	p-norm of x
$\mathcal{A}$	analysis function, definition 1.3, page 13
$A(D, R)$	analysis space, definition 1.1, page 12
$A_{\text{fuzzy}}(D, R)$	fuzzy analysis space, definition 1.8, page 18
J	objective function, definition 1.2, page 12
$\mathbb{B}$	the set of Boolean truth values {true, false}
c	number of clusters, $c = |K|$
$\delta_{i,j}$	Kronecker symbol, $\delta_{i,j} = 1$ for $i = j$, otherwise $\delta_{i,j} = 0$
D	data space, definition 1.1, page 12
R	result space, definition 1.1, page 12
K	cluster set, subset of the result space R
n	number of data, $n = |X|$
$\mathbb{N}$	natural numbers without zero
$\mathbb{N}_{\leq l}$	subset of the natural numbers, $\mathbb{N}_{\leq l} = \{1, 2, 3, \ldots, l\}$
$\mathbb{N}^*$	natural numbers including zero
$\mathbb{N}^*_{\leq l}$	subset of the natural numbers, $\mathbb{N}^*_{\leq l} = \{0, 1, 2, \ldots, l\}$
p	dimension (mostly the dimension of the data space)
$\mathcal{P}(A)$	power set of A, $\mathcal{P}(A) = \{X | X \subseteq A\}$
$\mathcal{P}_m(A)$	set of subsets of A with m elements
$\mathbb{R}$	real numbers
$\mathbb{R}_+, \mathbb{R}_{>0}$	positive real numbers
$\mathbb{R}_-, \mathbb{R}_{<0}$	negative real numbers
$\top$	$A^\top$ is the transposed matrix of A
$u_{i,j}$	membership of datum $x_j \in X$ in cluster $k_i \in K$
X	data set, subset of the data space D
$\Delta_{i,j}$	shortest distance vector from datum x_j to the contour of cluster k_i, page 199
α	parameter of the adaptive fuzzy clustering algorithm, page 70
η	extension factor for possibilistic clusters, pages 26, 48
τ	time index for the computational effort, page 58

respective scaling. When the formulae for the prototypes represent linear combinations of the data vectors (e.g. as in the fuzzy c-means algorithm), the analysis results are also invariant with respect to scaling.

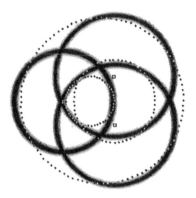

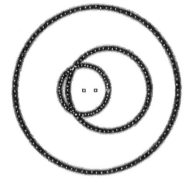

Figure A.1: FCSS analysis   Figure A.2: FCSS analysis

Algorithms that require a numerical iteration procedure to solve a system of non-linear equations are usually not invariant with respect to scaling of the data. If the distance function (or part of it) occurs quadratically, different gradients in the derivation result from scaling, especially for distances clearly less than 1 compared larger distances. Consequently, the algorithm may converge to a another (local) minimum for scaled data. But also for the algorithms which explicitly compute their prototypes the invariance can not be guaranteed. The fuzzy c-spherical shells algorithm has a high-degree non-linear distance function, which is also reflected in the formalae for the prototypes. Figures A.1 and A.2 show FCSS clustering results for differently scaled data sets. (The data sets were scaled to a uniform size for the figures.) Although the initialization of the prototypes was scaled accordingly, the results of the FCSS algorithm are different. In comparison to figure A.1 (largest circle radius 2), the data for the partition in figure A.2 were magnified by a factor 100 (largest circle radius 200). While the algorithm only needs 17 steps for figure A.1, several hundred iteration steps are computed for the intuitive partition shown in figure A.2.

A.3 Overview on FCQS cluster shapes

Each FCQS cluster is described by a quadric of the form $p_1 x_1^2 + p_2 x_2^2 + p_3 x_1 x_2 + p_4 x_1 + p_5 x_2 + p_6 = 0$. Defining

$$\Delta = \det \begin{pmatrix} p_1 & \frac{p_3}{2} & \frac{p_4}{2} \\ \frac{p_3}{2} & p_2 & \frac{p_5}{2} \\ \frac{p_4}{2} & \frac{p_5}{2} & p_6 \end{pmatrix} ,$$

$$I = p_1 + p_2,$$

$$J = \det \begin{pmatrix} p_1 & \frac{p_3}{2} \\ \frac{p_3}{2} & p_2 \end{pmatrix} \quad \text{and}$$

$$K = \det \begin{pmatrix} p_1 & \frac{p_4}{2} \\ \frac{p_4}{2} & p_6 \end{pmatrix} + \det \begin{pmatrix} p_2 & \frac{p_5}{2} \\ \frac{p_5}{2} & p_6 \end{pmatrix}$$

we obtain the quadric shapes from table A.3 (reference: [66], see also there for the three-dimensional case).

Δ	J	$\frac{\Delta}{I}$	K	Cluster shape
$\neq 0$	> 0	< 0		real ellipse
$\neq 0$	> 0	> 0		complex ellipse
$\neq 0$	< 0			hyperbola
$\neq 0$	0			parabola
0	< 0			two real, intersecting straight lines
0	> 0			conjugated complex, intersecting straight lines
0	0		< 0	two real, parallel straight lines
0	0		> 0	conjugated complex, parallel straight lines
0	0		0	coinciding straight lines

Table A.3: Two-dimensional quadrics

A.4 Transformation to straight lines

At the end of the FCQS algorithm, hyperbolas, double straight lines of long-stretched ellipses are transformed into linear clusters, since these shapes hardly occur in conventional applications but approximate straight line segments in practice. Considering the (rotated) quadric $p_1 x_1^2 + p_2 x_2^2 + p_4 x_1 + p_5 x_2 + p_6 = 0$ without mixed term $x_1 x_2$ (see section 4.7), Krishnapuram,

Frigui and Nasraoui describe in [66] the following transformation to straight lines $c_2 x_2 + c_1 x_1 + c_0 = 0$:

A hyperbola approximates two straight lines that roughly describe the position of their asymptotes. In this case, the straight line equations have the parameters $c_2 = 1$, $c_1 = \sqrt{\frac{p_1}{p_2}}$ and $c_0 = \frac{p_4}{2p_1}\sqrt{-\frac{p_1}{p_2}} + \frac{p_5}{2p_2}$ or $c_2 = 1$, $c_1 = \sqrt{-\frac{p_1}{p_2}}$ and $c_0 = -\frac{p_4}{2p_1}\sqrt{-\frac{p_1}{p_2}} + \frac{p_5}{2p_2}$, respectively. These are also the straight line equations for the case of two intersecting straight lines. Two parallel straight lines are parallel to one of the two main axes after rotation. The parallels to the horizontal axis (i.e. $p_1 \approx p_4 \approx 0$) result from $c_2 = 1$, $c_1 = 0$ and $c_0 = \frac{p_5 \pm \sqrt{p_5^2 - 4p_2 p_6}}{2p_2}$, the parallels to the vertical axis (i.e. $p_2 \approx p_5 \approx 0$) from $c_2 = 0$, $c_1 = 1$ and $c_0 = \frac{p_4 \pm \sqrt{p_4^2 - 4p_1 p_6}}{2p_1}$. Sometimes, a straight line is approximated by a long-stretched ellipse. We recognize a long-stretched ellipse by the quotient of the longer and the shorter ellipse's axes (cf. remark 7.16) which is greater than a certain threshold C (≈ 10). Similar to the case of parallel straight lines, the ellipse was rotated in such a way that its axes are parallel to the main axes. We obtain the resulting straight lines analogously to the previous case. We can also cover the case of a flat hyperbola in a similar way: again, we recognize their degree of stretching computing the quotient of their half axes: If the cross axis is parallel to the horizontal axis, i.e. $\frac{1}{p_2}\left(\frac{p_4^2}{4p_1} + \frac{p_5^2}{4p_2} - p_6\right) < 0$, it is a flat hyperbola when $-\frac{p_1}{p_2} > C^2$ holds, and in the other case $-\frac{p_2}{p_1} > C^2$. Once more, we receive the straight line equations analogously to the case of parallel straight lines. After the computation of the straight line parameters, the straight lines have to be rotated back into the original coordinate system.

References

[1] G.P. Babu, M.N. Murty. *Clustering with Evolutionary Strategies.* Pattern Recognition 27 (1994), 321-329

[2] R. Babuška, H.B. Verbruggen. *A New Identification Method for Linguistic Fuzzy Models.* Proc. Intern. Joint Conferences of the Fourth IEEE Intern. Conf. on Fuzzy Systems and the Second Intern. Fuzzy Engineering Symposium, Yokohama (1995), 905-912

[3] J. Bacher. *Clusteranalyse.* Oldenbourg, München (1994)

[4] T. Bäck. *Evolutionary Algorithms in Theory and Practice.* Oxford University Press, Oxford (1996).

[5] H. Bandemer, S. Gottwald. *Einführung in Fuzzy-Methoden.* (4. Aufl.), Akademie Verlag, Berlin (1993)

[6] H. Bandemer, W. Näther. *Fuzzy Data Analysis.* Kluwer, Dordrecht (1992)

[7] M. Barni, V. Cappellini, A. Mecocci. *Comments on "A Possibilistic Approach to Clustering".* IEEE Trans. Fuzzy Systems 4 (1996), 393-396

[8] J.C. Bezdek. *Fuzzy Mathematics in Pattern Classification.* Ph. D. Thesis, Applied Math. Center, Cornell University, Ithaca (1973)

[9] J.C. Bezdek. *A Convergence Theorem for the Fuzzy ISODATA Clustering Algorithms.* IEEE Trans. Pattern Analysis and Machine Intelligence 2 (1980), 1-8

[10] J.C. Bezdek. *Pattern Recognition with Fuzzy Objective Function Algorithms.* Plenum Press, New York (1981)

[11] J.C. Bezdek, C. Coray, R. Gunderson, J. Watson. *Detection and Characterization of Cluster Substructure.* SIAM Journ. Appl. Math. 40 (1981), 339-372

[12] J.C. Bezdek, R.H. Hathaway. *Numerical Convergence and Interpretation of the Fuzzy C-Shells Clustering Algorithm.* IEEE Trans. Neural Networks 3 (1992), 787-793

[13] J.C. Bezdek, R.J. Hathaway, N.R. Pal. *Norm-Induced Shell-Prototypes (NISP) Clustering.* Neural, Parallel & Scientific Computations 3 (1995), 431-450

[14] J.C. Bezdek, R.H. Hathaway, M.J. Sabin, W.T. Tucker. *Convergence Theory for Fuzzy c-Means: Counterexamples and Repairs.* IEEE Trans. Systems, Man, and Cybernetics 17 (1987), 873-877

[15] J.C. Bezdek, N.R. Pal. *Two Soft Relatives of Learning Vector Quantization.* Neural Networks 8 (1995), 729-743

[16] H.H. Bock. *Automatische Klassifikation.* Vandenhoeck & Ruprecht, Göttingen (1974)

[17] H.H. Bock. *Clusteranalyse mit unscharfen Partitionen.* In: H.H. Bock (ed.). Klassifikation und Erkenntnis: Vol. III: Numerische klassifikation. INDEKS, Frankfurt (1979), 137-163

[18] H.H. Bock. *Classification and Clustering: Problems for the Future.* In: E. Diday, Y. Lechevallier, M. Schrader, P. Bertrand, B. Burtschy (eds.). *New Approaches in Classification and Data Analysis.* Springer, Berlin(1994), 3-24

[19] H.H. Bock. *Probability Models and Hypotheses Testing in Partitioning Cluster Analysis.* In: P. Arabic, L.J. Hubert, G. De Soete (eds.). *Clustering and Classification.* World Scientific, River Edge, NJ (1996), 379-453

[20] T.W. Cheng, D.B. Goldgof, L.O. Hall. *Fast Clustering with Application to Fuzzy Rule Generation.* Proc. IEEE Intern. Conf. on Fuzzy Systems, Yokohama (1995), 2289-2295

[21] T.W. Cheng, D.B. Goldgof, L.O. Hall. *Fast Fuzzy Clustering.* Fuzzy Sets and Systems 93 (1998), 49-56

[22] R.N. Davé. *Use of the Adaptive Fuzzy Clustering Algorithm to Detect Lines in Digital Images.* Proc. Intelligent Robots and Computer Vision VIII, Vol. 1192, (1989), 600-611

[23] R.N. Davé. *Fuzzy Shell-Clustering and Application To Circle Detection in Digital Images*. Intern. Journ. General Systems 16 (1990), 343-355

[24] R.N. Davé. *New Measures for Evaluating Fuzzy Partitions Induced Through c-Shells Clustering*. Proc. SPIE Conf. Intell. Robot Computer Vision X, Vol. 1607 (1991), 406-414

[25] R.N. Davé. *Characterization and Detection of Noise in Clustering*. Pattern Recognition Letters 12 (1991), 657-664

[26] R.N. Davé. *On Generalizing Noise Clustering Algorithms*. In: Proc. 7th Intern. Fuzzy Systems Association World Congress (IFSA'97) Vol. III, Academia, Prague (1997), 205-210

[27] R.N. Davé, K. Bhaswan. *Adaptive Fuzzy c-Shells Clustering and Detection of Ellipses*. IEEE Trans. Neural Networks 3 (1992), 643-662

[28] R.N. Davé, R. Krishnapuram. *Robust Clustering Methods: A Unified View*. IEEE Trans. Fuzzy Systems 5 (1997), 270-293

[29] D. Driankov, H. Hellendoorn, M. Reinfrank. *An Introduction to Fuzzy Control*. Springer, Berlin (1993)

[30] R. Duda, P. Hart. *Pattern Classification and Scene Analysis*. Wiley, New York (1973)

[31] J.C. Dunn. *A Fuzzy Relative of the ISODATA Process and its Use in Detecting Compact, Well separated Clusters*. Journ. Cybern. 3 (1974), 95-104

[32] G. Engeln-Müllges, F. Reutter. *Formelsammlung zur Numerischen Mathematik mit Turbo Pascal Programmen*. BI Wissenschaftsverlag, Wien (1991)

[33] U.M. Fayyad, G. Piatetsky-Shapiro, P. Smyth, R. Uthurusamy. *Advances in Knowledge Discovery and Data Mining*. AAAI Press, Menlo Park (1996)

[34] D.P. Filev, R.R. Yager. *A Generalized Defuzzification Method via BAD Distributions*. Intern. Journ. of Intelligent Systems 6 (1991), 687-697

[35] H. Frigui, R. Krishnapuram. *On Fuzzy Algorithms for the Detection of Ellipsoidal Shell Clusters*. Technical Report, Department of Electrical and Computer Engineering, University of Missouri, Columbia

[36] H. Frigui, R. Krishnapuram. *A Comparison of Fuzzy Shell-Clustering Methods for the Detection of Ellipses*. IEEE Trans. Fuzzy Systems 4 (1996), 193-199

[37] I. Gath, A.B. Geva. *Unsupervised Optimal Fuzzy Clustering*. IEEE Trans. Pattern Analysis and Machine Intelligence 11 (1989), 773-781.

[38] I. Gath, D. Hoory. *Fuzzy Clustering of Elliptic Ring-Shaped Clusters*. Pattern Recognition Letters 16 (1995), 727-741

[39] H. Genther, M. Glesner. *Automatic Generation of a Fuzzy Classification System Using Fuzzy Clustering Methods*. Proc. ACM Symposium on Applied Computing (SAC'94), Phoenix (1994), 180-183

[40] H.M. Groscurth, K.P. Kress. *Fuzzy Data Compression for Energy Optimization Models*. Energy 23 (1998), 1-9

[41] E.E. Gustafson, W.C. Kessel. *Fuzzy Clustering with a Fuzzy Covariance Matrix*. IEEE CDC, San Diego, California (1979), 761-766

[42] R.J. Hathaway, J.C. Bezdek. *Optimization of Clustering Criteria by Reformulation*. IEEE Trans. on Fuzzy Systems 3 (1995), 241-245

[43] F. Höppner. *Fuzzy Shell Clustering Algorithms in Image Processing: Fuzzy C-Rectangular and 2-Rectangular Shells*. IEEE Trans. on Fuzzy Systems 5 (1997), 599-613

[44] P.V.C. Hough. *Method and Means for Recognizing Complex Patterns*. U.S. Patent 3069654, 1962

[45] J. Illingworth, J. Kittler. *The Adaptive Hough Transform*. IEEE Trans. Pattern Analysis and Machine Intelligence 9 (1987), 690-698

[46] D. Karen, D. Cooper, J. Subrahmonia. *Describing Complicated Objects by Implizit Polynomials*. IEEE Trans. Pattern Analysis and Machine Intelligence 16 (1994), 38-53

[47] M.S. Kamel, S.Z. Selim. *A Thresholded Fuzzy c-Means Algorithm for Semi-Fuzzy Clustering*. Pattern Recognition 24 (1991), 825-833

[48] A. Kaufmann, M.M. Gupta. *Introduction to Fuzzy Arithmetic*. Reinhold, New York (1985)

[49] U. Kaymak, R. Babuška. *Compatible Cluster Merging for Fuzzy Modelling*. Proc. Intern. Joint Conferences of the Fourth IEEE Intern. Conf. on Fuzzy Systems and the Second Intern. Fuzzy Engineering Symposium, Yokohama (1995), 897-904

[50] F. Klawonn. *Fuzzy Sets and Vague Environments.* Fuzzy Sets and Systems 66 (1994), 207-221

[51] F. Klawonn. *Similarity Based Reasoning.* Proc. Third European Congress on Intelligent Techniques and Soft Computing (EUFIT'95), Aachen (1995), 34-38

[52] F. Klawonn, J. Gebhardt, R. Kruse. *Fuzzy Control on the Basis of Equality Relations - with an Example from Idle Speed Control.* IEEE Trans. Fuzzy Systems 3 (1995), 336-350

[53] F. Klawonn, A. Keller. *Fuzzy Clustering and Fuzzy Rules.* Proc. 7th International Fuzzy Systems Association World Congress (IFSA'97) Vol. I, Academia, Prague (1997), 193-198

[54] F. Klawonn, A. Keller. *Fuzzy Clustering with Evolutionary Algorithms.* Intern. Journ. of Intelligent Systems 13 (1998), 975-991

[55] F. Klawonn, E.-P. Klement. *Mathematical Analysis of Fuzzy Classifiers.* In: X. Liu, P. Cohen, M. Berthold (eds.). *Advances in Intelligent Data Analysis.* Springer, Berlin (1997), 359-370

[56] F. Klawonn, R. Kruse. *Automatic Generation of Fuzzy Controllers by Fuzzy Clustering.* Proc. 1995 IEEE Intern. Conf. on Systems, Man and Cybernetics, Vancouver (1995), 2040-2045

[57] F. Klawonn, R. Kruse. *Derivation of Fuzzy Classification Rules from Multidimensional Data.* In: G.E. Lasker, X. Liu (eds.). Advances in Intelligent Data Analysis. The International Institute for Advanced Studies in Systems Research and Cybernetics, Windsor, Ontario (1995), 90-94

[58] F. Klawonn, R. Kruse. *Constructing a Fuzzy Controller from Data.* Fuzzy Sets and Systems 85 (1997), 177–193

[59] G.J. Klir, B. Yuan. *Fuzzy Sets and Fuzzy Logic.* Prentice Hall, Upper Saddle River, NJ (1995)

[60] R. Krishnapuram, L.F. Chen. *Implementation of Parallel Thinning Algorithms Using Recurrent Neural Networks.* IEEE Trans. Neural Networks 4 (1993), 142-147

[61] R. Krishnapuram, C.P. Freg. *Fitting an Unknown Number of Lines and Planes to Image Data Through Compatible Cluster Merging.* Pattern Recognition 25 (1992), 385-400

[62] R. Krishnapuram, J. Keller. *A Possibilistic Approach to Clustering.* IEEE Trans. Fuzzy Systems 1 (1993), 98-110

[63] R. Krishnapuram, J. Keller. *The Possibilistic C-Means Algorithm: Insights and Recommendations.* IEEE Trans. Fuzzy Systems 4 (1996), 385-393

[64] R. Krishnapuram, H. Frigui, O. Nasraoui. *New Fuzzy Shell Clustering Algorithms for Boundary Detection and Pattern Recognition.* Proceedings of the SPIE Conf. on Intelligent Robots and Computer Vision X: Algorithms and Techniques, Boston (1991), 458-465

[65] R. Krishnapuram, H. Frigui, O. Nasraoui. *The Fuzzy C Quadric Shell Clustering Algorithm and the Detection of Second-Degree Curves.* Pattern Recognition Letters 14 (1993), 545-552

[66] R. Krishnapuram, H. Frigui, O. Nasraoui. *Fuzzy and Possibilistic Shell Clustering Algorithms and Their Application to Boundary Detection and Surface Approximation – Part 1 & 2.* IEEE Trans. Fuzzy Systems 3 (1995), 29-60

[67] R. Krishnapuram, O. Nasraoui, H. Frigui. *The Fuzzy C Spherical Shells Algorithms: A New Approach.* IEEE Trans. Neural Networks 3 (1992), 663-671.

[68] R. Kruse, F. Klawonn, J. Gebhardt. *Foundations of Fuzzy Systems.* Wiley, Chichester (1994)

[69] R. Kruse, K.D. Meyer. *Statistics with Vague Data.* Reidel, Dordrecht (1987)

[70] L.I. Kuncheva. *Editing for the k-Nearest Neighbors Rule by a Genetic Algorithm.* Pattern Recognition Letters 16 (1995), 809-814

[71] E.H. Mamdani, S. Assilian. *An Experiment in Linguistic Synthesis with a Fuzzy Logic Controller.* Intern. Journ. of Man-Machine Studies 7 (1975), 1-13

[72] Y. Man, I. Gath. *Detection and Separation of Ring-Shaped Clusters Using Fuzzy Clustering.* IEEE Trans. Pattern Analysis and Machine Intelligence 16 (1994), 855-861

[73] K.G. Manton, M.A. Woodbury, H.D. Tolley. *Statistical Applications Using Fuzzy Sets.* Wiley, New York (1994)

[74] J.J. More. *The Levenberg-Marquardt Algorithm: Implementation and Theory.* In: A. Dold, B. Eckmann (eds.). Numerical Analysis, Springer, Berlin (1977), 105-116

[75] Z. Michalewicz. *Genetic Algorithms + Data Structures = Evolution Programs.* Springer, Berlin (1992).

[76] Y. Nakamori, M. Ryoke. *Identification of Fuzzy Prediction Models Through Hyperellipsoidal Clustering.* IEEE Trans. Systems, Man, and Cybernetics 24 (1994), 1153-1173

[77] H. Narazaki, A.L. Ralescu. *An Improved Synthesis Method for Multilayered Neural Networks Using Qualitative Knowledge.* IEEE Trans. on Fuzzy Systems 1(1993), 125-137

[78] D. Nauck, F. Klawonn. *Neuro-Fuzzy Classification Initialized by Fuzzy Clustering.* Proc. of the Fourth European Congress on Intelligent Techniques and Soft Computing (EUFIT'96), Aachen (1996), 1551-1555

[79] D. Nauck, F. Klawonn, R. Kruse. *Neuro-Fuzzy Systems.* Wiley, Chichester (1997)

[80] Y. Ohashi. *Fuzzy Clustering and Robust Estimation.* 9th Meeting SAS User Grp. Int., Hollywood Beach, Florida (1984)

[81] N.R. Pal, K. Pal, J.C. Bezdek, T.A. Runkler. *Some Issues in System Identification Using Clustering.* Proc. IEEE Intern. Conf. on Neural Networks, Houston (1997), 2524-2529

[82] N. Pfluger, J. Yen, R. Langari. *A Defuzzification Strategy for a Fuzzy Logic Controller Employing Prohibitive Information in Command Formulation.* Proc. IEEE Intern. Conf. on Fuzzy Systems, San Diego (1992), 717-723

[83] V. Pratt. *Direct Least Squares Fitting on Algebraic Surfaces.* Computer Graphics 21 (1987), 145-152

[84] T.A. Runkler. *Selection of Appropriate Defuzzification Methods Using Application Specific Properties.* IEEE Trans. on Fuzzy Systems 5 (1997), 72-79

[85] T.A. Runkler. *Automatic Generation of First Order Takagi-Sugeno Systems Using Fuzzy c-Elliptotypes Clustering.* Journal of Intelligent and Fuzzy Systems 6 (1998), 435-445

[86] T.A. Runkler, J.C. Bezdek. *Living Clusters: An Application of the El Farol Algorithm to the Fuzzy c-Means Model.* Proc. European Congress on Intelligent Techniques and Soft Computing, Aachen (1997), 1678-1682

[87] T.A. Runkler, J.C. Bezdek. *Polynomial Membership Functions for Smooth First Order Takagi-Sugeno Systems.* In A. Grauel, W. Becker, F. Belli (eds.). Fuzzy-Neuro-Systeme '97, volume 5 of Proceedings in Artificial Intelligence. Soest (1997), 382-388

[88] T.A. Runkler, J.C. Bezdek. *Function Approximation with Polynomial Membership Functions and Alternating Cluster Estimation.* Fuzzy Sets and Systems 101 (1999)

[89] T.A. Runkler, J.C. Bezdek. *RACE: Relational Alternating Cluster Estimation and the Wedding Table Problem.* In W. Brauer (ed.). Fuzzy-Neuro-Systems '98, volume 7 of Proceedings in Artificial Intelligence. München (1998) 330-337

[90] T.A. Runkler, J.C. Bezdek. *Regular Alternating Cluster Estimation.* Proc. European Congress on Intelligent Techniques and Soft Computing. Aachen,(1998), 1355-1359

[91] T.A. Runkler, J.C. Bezdek. *Alternating Cluster Estimation: A New Tool for Clustering and Function Approximation.* IEEE Trans. on Fuzzy Systems (to appear 1999)

[92] T.A. Runkler, J. Hollatz, H. Furumoto, E. Jünnemann, K. Villforth. *Compression of Industrial Process Data Using Fast Alternating Cluster Estimation (FACE).* In: GI Jahrestagung Informatik (Workshop Data Mining), Magdeburg (1998), 71-80

[93] T.A. Runkler, R.H. Palm. *Identification of Nonlinear Systems Using Regular Fuzzy c-Elliptotype Clustering.* Proc. IEEE Intern. Conf. on Fuzzy Systems, New Orleans (1996), 1026-1030

[94] B. Schweizer, A. Sklar. *Associative Functions and Statistical Triangle Inequalities.* Publicationes Mathematicae Debrecen 8 (1961), 169-186

[95] P. Spellucci. *Numerische Verfahren der nichtlinearen Optimierung.* Birkhäuser Verlag (1993)

[96] D. Steinhausen, K. Langer. *Clusteranalyse.* deGruyter, Berlin (1977)

[97] M. Sugeno, T. Yasukawa. *A Fuzzy-Logic-Based Approach to Qualitative Modeling.* IEEE Trans. Fuzzy Systems 1 (1993), 7-31

[98] T. Takagi, M. Sugeno. *Fuzzy Identification of Systems and its Application to Modeling and Control.* IEEE Trans. on Systems, Man, and Cybernetics 15 (1985), 116-132

[99] M.P. Windham. *Cluster Validity for Fuzzy Clustering Algorithms.* Fuzzy Sets and Systems 3 (1980), 1-9

[100] M.P. Windham. *Cluster Validity for the Fuzzy c-Means Clustering Algorithms.* IEEE Trans. Pattern Analysis and Machine Intelligence 11 (1982), 357-363

[101] X.L. Xie, G. Beni. *A Validity Measure for Fuzzy Clustering.* IEEE Trans. Pattern Analysis and Machine Intelligence 13 (1991), 841-847

[102] R.R. Yager, D.P. Filev. *SLIDE: A Simple Adaptive Defuzzification Method.* IEEE Trans. on Fuzzy Systems 1 (1993), 69-78

[103] Y. Yoshinari, W. Pedrycz, K. Hirota. *Construction of Fuzzy Models Through Clustering Techniques.* Fuzzy Sets and Systems 54 (1993), 157-165

[104] L.A. Zadeh. *Fuzzy Sets.* Information and Control 8 (1965), 338-353

Index

Printed and bound by CPI Group (UK) Ltd, Croydon, CR0 4YY

27/10/2024

14580206-0002